Mathematical Concepts
in
Geometrical Physics

Mathematical Concepts in Geometrical Physics

Josef Carter

Oxford Book Company
Jaipur ▪ New Delhi

Published By

Oxford Book Company

Regd. Off. "Bony Residency", Gate No. 2
Opp. Tilak Public School, Vishweswariya Nagar,
Gopalpura Road, Jaipur - 302018, India
Ph: 0141 - 2761280, Telefax: 2761381
email - oxfordbookjpr@gmail.com
website - oxfordbookcompany.com

Delhi Office: 102, 1st Floor, Satyam House,
4327/3, Ansari Road, Darya Ganj,
New Delhi - 110002, India
Ph: 011-45652440

Distributed by
KEYA INTERNATIONAL PRESS LLP
E-mail: keyainternationalpressllp@gmail.com

First published in India, 2024

ISBN : 9789355240620

Typeset by Shivangi Computers, Jaipur

Printed and bound in India by Replika Press Pvt. Ltd.

Contents

Preface

Physicists have discovered a jewel-like geometric object that dramatically simplifies calculations of particle interactions and challenges the notion that space and time are fundamental components of reality. The revelation that particle interactions, the most basic events in nature, may be consequences of geometry significantly advances a decades-long effort to reformulate quantum field theory, the body of laws describing elementary particles and their interactions. Interactions that were previously calculated with mathematical formulas thousands of terms long can now be described by computing the volume of the corresponding jewel-like "amplituhedron," which yields an equivalent one-term expression. The new geometric version of quantum field theory could also facilitate the search for a theory of quantum gravity that would seamlessly connect the large- and small-scale pictures of the universe. Attempts thus far to incorporate gravity into the laws of physics at the quantum scale have run up against nonsensical infinities and deep paradoxes. The amplituhedron, or a similar geometric object, could help by removing two deeply rooted principles of physics: locality and unitarity.

Locality is the notion that particles can interact only from adjoining positions in space and time. And unitarity holds that the probabilities of all possible outcomes of a quantum mechanical interaction must add up to one. The concepts are the central pillars of quantum field theory in its original form, but in certain situations involving gravity, both break down, suggesting neither is a fundamental aspect of nature. In keeping with this idea, the new geometric approach to particle interactions removes locality and unitarity from its starting assumptions. The amplituhedron is not built out of space-time and probabilities; these properties merely arise as consequences of the jewel's geometry. The usual picture of space and time, and particles moving around in them, is a construct. The 60-year-old method for calculating scattering amplitudes — a major innovation at the time — was pioneered by the Nobel Prize-winning physicist Richard Feynman. He sketched line drawings of all the ways a scattering process could occur and then summed the likelihoods of the different drawings. The simplest Feynman diagrams look like trees: The particles involved in a collision come together like roots, and the particles that result shoot out like branches. More complicated diagrams have loops, where colliding particles turn into unobservable "virtual particles" that interact with each other before branching out as real final products. There are diagrams with one loop, two loops, three loops and so on — increasingly baroque iterations of the scattering process that contribute progressively less to its total amplitude. Virtual particles are never observed in nature, but they were considered mathematically necessary for unitarity — the requirement that probabilities sum to one.

The book's goal is to present mathematical ideas associated with geometrical physics in a rather introductory language. Included are many examples from elementary physics and also, for those wishing to reach a higher level of understanding, a more advanced treatment of the mathematical topics. It is aimed as an elementary text, more so than most others on the market, and is intended for graduate students.

— Josef Carter

Chapter 1

Systems of Equations and Inequalities

Solving Systems of Equations

Up until this point, we have been dealing with only one equation at a time. Now, we will work with more than variable and more than one equation. These are called systems of equations. When answering a system of equations, you need to give the value for each variable.

Solving Systems of Linear Equations

When we're through covering the two chapters on solving systems of equations, there will be six ways that we can use to solve a system of linear equations

Graphically: Graph both equations and find the intersection point.

Inaccurate by hand.

Useful when using technology.

More appropriate for non-linear systems.

Must solve for the equation for y first.

Substitution: Solve one equation for one variable and then substitute that into the other equation.

Best algebraic technique for non-linear systems.

Works well when a variable can be solved for easily, has a coefficient of one.

Works better when fractions and roots aren't involved.

Addition / Elimination: Multiply one or more equations by a constant and then add the two equations together to eliminate one variable.

Works well for a linear system when there is no variable with a coefficient of one.

Works well for 2x2 (2 equations with 2 variables) systems of equations, but becomes tedious and labour intensive for larger systems.

Gaussian Elimination / Gauss Jordan Elimination: Uses elementary operations to produce equivalent equations.

Works for non-square systems of linear equations.

Built upon the concepts of addition elimination, but instead of obtaining new equations, the old equation is replaced with an equivalent equation.

When applied with matrices, probably the fastest way to solve a large system of linear equations by hand. Certainly the instructor's favourite method.

Cramer's Rule: Uses determinants of a matrix to find the solution.

Works only for square systems of linear equations where the determinant of the coefficient matrix isn't zero.

Good for a computer or calculator where there is a determinant program.

Slow by hand.

Slow on the calculator without a program since every determinant must be entered manually.

Can be used when you need to find just one of the variables.

Matrix Algebra / Matrix Inverses: Uses the inverse of a matrix to find the solution.

Works only for square systems of linear equations where the determinant of the coefficient matrix isn't zero.

Good for a computer or calculator where there is a matrix inverse function.

Slow by hand.

Quick to do on the calculator.

Will return decimal answers, but you can use the fraction key to convert it to integers.

Substitution

The method of substitution will work with non-linear as well as linear equations.

1. Solve one of the equations for one of the variables.
2. Substitute that expression in for the variable in the other equation.
3. Solve the equation for the remaining variable
4. Back-substitute the value for the variable to find the other variable.
5. Check.

The process of back-substitution involves taking the value of the variable found in step 3 and substituting it back into the expression obtained in step 1 (or the original problem) to find the remaining variable. It is important that both variables be given when solving a system of equations.

A common mistake students make is to find one variable and stop there. You need to include a value for all the variables. It is a good idea to check your answer into the both equations, but is probably sufficient to check in the equation you didn't isolate a variable in the first step. That is, if you solved for y in the first equation in step 1, use the second equation to check the answer.

Graphical Approach

The graphical approach works well with a graphing calculator, but is inaccurate by hand (did those points intersect at 1/6 or 1/7?) unless the graph happens to fall exactly on the grid lines.

1. Solve each equation for y. This may involve a plus and minus if there is a y^2 term. If you're not graphing with a calculator or computer, you can skip this step.
2. Graph each equation.
3. Find the points of intersection.
4. Check!

It is important to check your answers to make sure that you have read the intersection point correctly. Sometimes the calculator will fail to give an intersection point using the intersect command. You may need to use the trace feature of the calculator to find the intersection point. You may use your calculator to check the answer. Try to convert your answer to fractional form if possible. The graphical approach can save a lot of time when you're working with a non-linear system of equations.

Systems of Linear Equations in Two Variables

Addition / Elimination

The idea behind the addition / elimination method is to multiple one or more equations by a constant so when they are added together, one of the variables eliminates. Then you have one equation with one variable and you can solve for that variable.

1. Choose a variable to eliminate. Usually the variable that can be eliminated by multiplying by smaller numbers is the better choice.
2. Multiply one or both equations by a constant so that the least common multiple of the coefficients on the variable to be eliminated is obtained. Care should be taken so that one coefficient becomes negative and the other is positive.
3. Add the two equations together so the variable is eliminated.
4. Solve the resulting equation for the remaining variable.
5. Back-substitute that value into the one of the two original equations to find the remaining variable.
6. Check your answer into the other equation.

As an alternative to step 5, and this is extremely helpful when the answer is a fraction or decimal value and not pleasant to work with, you can go through the elimination process again with the other variable, and then you don't have to work with the fractions until the check process.

Graphical Interpretation of Solutions

You may have either no solution, one unique solution, or many solutions when solving a 2x2 system of linear equations.

Exactly One Solution

- Intersecting lines
- Consistent System (consistent means there is a solution, there is no contradiction).
- Independent System (the value for y doesn't depend on what x is).

Write your answer as the set containing the ordered pair {(x,y)} (replace the x and y with the actual values).

This would be analogous to a conditional system from section 2.1, certain values of x and y make it true, while others don't.

Consider the system of linear equations $3x + 2y = 17$ and $2x - y = 2$. The solution is $x = 3$ and $y = 4$, so you write the answer as $\{ (3, 4) \}$.

No Solution

- Parallel lines
- Inconsistent System (contradiction)
- Independent / Dependent doesn't apply since there is no solution.

Write your answer as "no solution", the symbol for the empty or null set, Ø, or the empty set { }, but do not write it as the set containing the null set {Ø}. Be careful if you're a person who slashes their zeros. Make sure I can tell a zero apart from the null set.

This case occurs when both variables eliminate, and you are left with a false statement.

This would be analogous to a contradiction from section 2.1.

Consider the system of linear equations $3x - 2y = 3$ and $-3x + 2y = 2$. When you add them together, you get $0 = 5$, which is a contradiction. Therefore, the answer is "no solution", { }, or Ø

Many Solutions

- Coincident lines (same line)
- Consistent System (identity)
- Dependent System (the value for y will depend on what x is, it's not always the same value).

Write your answer as one of the two equations. Do NOT say "many solutions" or "all real numbers". All real numbers will not work. First of all, it the solutions are ordered pairs, not individual x or y coordinates. Second, not every point works, only those on the line work.

You may also write your answer in parametric form. This will be the preferred method for higher ordered systems, so you might as well learn it now.

This case occurs when both variables eliminate, and you are left with a true statement.

This would be analogous to an identity from section 2.1.

Consider the system of linear equations $3x - 2y = 3$ and $-3x + 2y = -3$. When you add them together, you get $0 = 0$, which is always true. Therefore, any values of x and y that satisfy the equation are solutions. You would write your solution as:

- 3x - 2y = 3
- { (x, y) | 3x - 2y = 3 }
- x = t, y = 3/2 t - 3/2 (this is known as parametric form and is covered in section 8.3)

Making "Nice" Problems

Ever wonder why most all of the problems in an algebra textbook come out with nice integer answers? Is it because life is that way? Certainly not, it is because the problems are contrived to have nice answers.

How do you get a made up problem to have nice answers? The answer is that you don't. You start with the answer and work backwards.

Let's say that we want the answer to be (3,-2).

Make up anything for the left side of the equation, say 2x-5y. Then, plug in x=3 and y=-2, so 2(3) - 5(-2) = 6 + 10 = 16. The first equation is 2x-5y=16.

Now, repeat the process again with a different left side, say 3x+2y. Well, 3(3) + 2(-2) = 9 - 4 = 5, so 3x+2y=5.

Your system of linear equations is 2x-5y=16 and 3x+2y=5.

You can make up anything for the left hand side, just make sure the right hand side is what you get when you plug those values in. You will also need to have two equations if there are two variables, one equation if there is only one variable, three equations if there are three variables, etc.

Least Squares Regression Model

Up until this point, when we have found the linear regression model, we have just used the functions on the calculator to obtain the results, and it has been fairly easy and painless. Now we learn that the calculator is actually solving a system of linear equations to obtain the model.

Summation Notation

The capital Greek letter sigma stands for sum. Normally, there is an index with a starting point (k=1) written below the sigma and an ending point (n, meaning k=n) written above the sigma. Then, each variable will have a subscript to let you know that it is a function or sequence that depends on the value of the index, k.

Confused? Well, maybe you should be. The book doesn't discuss sequences and series, they expect you to know how to do it.

I'm going to use a short hand notation to keep things simpler and easier to remember.

Instead of writing it the way the book does, I'll just use a sigma symbol and then what I want to total.

$$\sum_{k=1}^{n} x_k = \sum x$$

Remember that sigma means sum, so sigma x means add up all the x's.

In statistics, we like to simplify things, and get a little sloppy, and drop all the index stuff and just know that it applies to all the points. In the notation above, I will use the form on the right. "x just means add up all the x values. Not too bad when you look at it that way.

Linear Regression

Consider the linear model y = ax + b. The values for a and b can be found by solving this system of linear equations.

$$b\Sigma 1 + a\Sigma x = -y$$

$$b\Sigma x + a\Sigma x^2 = -xy$$

Notice each term in the second equation has one more x in it than the corresponding term in the first equation. This pattern will repeat when we do quadratic regression, or cubic, or quartic, or...

You will add up each variable in the summation for each different point. The first summation is the sum of 1.

So, if you add 1 for each point, you will simply have the number of points. The other values are the sum of the x's, the sum of the y's, the sum of the squares of the x's, and the sum of the products of the x's and y's.

Writing these ordered pairs in a columnar table, and then adding columns for the x^2 and xy will help. After you solve the system of linear equations, substitute the values for a and b into the equation y = ax + b to get the model.

Find the equation of the linear model that best fits the points (2,3), (5,2), (6,1), and (8,-1).

Set up a table with columns for x, y, x^2, and xy.

x	y	x^2	xy
2	3	4	6
5	2	25	10
6	1	36	6
8	-1	64	-8
21	5	129	14

The numbers in the bottom row represent the sums that go into the system. Since there are 4 points, the −1 = 4. The system of linear equations to solve is 4b + 21a = 5 and 21b + 129a = 14. When you solve that, you get a = -49/75 and b = 117/25. When you stick those back into the model, you get the y = -49/75x + 117/25.

Multivariable Linear Systems

Row-Echelon Form

Row-echelon form is basically a stair-step pattern with leading coefficients of one. When a system is placed into row-echelon form, back substitution is very easy. The bottom row gives the answer for z. That answer is back-substituted into the second equation and y is found. Then both y and z are substituted into the first equation and x is found.

Example: Use back substitution to solve the following system.

$$x - y + 2z = 5$$
$$y - z = -1$$
$$z = 3$$

1. The bottom equation gives us that z = 3.
2. Plugging z = 3 into the second equation gives y - 3 = -1 or y = 2.
3. Plugging y = 2 and z = 3 into the first equation gives x - 2 + 2(3) = 5.
4. Solving for x gives 1, so the solution is { (1, 2, 3) }.

Gaussian Elimination

Gaussian Elimination is named after Carl Friedrich Gauss, the German mathematician who proved the fundamental theorem of algebra. Two systems of equations are *equivalent* if they have the same solution set.

Elementary Operations

There are three basic operations, called elementary operations, that can be performed and that will render an equivalent system.

- Interchange two equations.
- Multiply one equation by a non-zero constant.
- Multiply an equation by a non-zero constant and add it to another equation, replacing that equation.

The addition / elimination technique that we discussed in section 5.2 uses these operations, they just weren't formalized. In the elimination technique, we could switch the two equations around and it wouldn't affect the solution set. We could multiply one or both equations by a non-zero constant and then add them together. There is nothing wrong with multiplying both equations by a non-zero constant, just that what's given here is the elementary or basic operations and multiplying both by a non-zero constant and adding would be a combination of these.

The Number of Solutions to a Linear System

Just like with a 2×2 system of linear equations, a larger system can have one, none, or many solutions.

Unique Solution: A consistent and independent system. The solution is given as an ordered triplet.

No Solution: An inconsistent system. There is no solution, and the answer should be written as the null set or the empty set, but not the set containing the null set. This results if at any point while trying to solve the system, an equivalent form has a contradiction (variables eliminate and the statement is false).

Many Solutions: A consistent and dependent system. The solution is given in parametric form. This case results when there are less equations than variables and no contradictions.

Example: Solve the following system of linear equations.

$3x - 2y + z = 1$

$y + 2z = 3$

1. Take the second equation and solve for y to get $y = 3 - 2z$.
2. Substitute $y = 3 - 2z$ into the first equation for y. This gives you $3x - 2(3 - 2z) + z = 1$, which is an equation with two variables, x and z. A little bit of simplification givs $3x - 6 + 4z + z = 1$ or $3x + 5z = 7$.

3. Since you earlier solved for y in terms of z, we now solve for x in terms of z. It is important that we solve all other variables in terms of the same variable. Solving $3x + 5z = 7$ for x gives $x = (7 - 5z) / 3$
4. Give the solution. Since x and y are both in terms of a third variable, z, we let $z = t$ and get:
   - $x = 1/3\ (7 - 5t)$
   - $y = 3 - 2t$
   - $z = t$

Mathematical Models

If you have the right number of ordered pairs, you can fit a model by solving a system a system of linear equations. To have an unique solution, you need as many points as constants.

Linear Model: $y=Ax+B$

There are two values that need to be found, A and B. Therefore, it takes two points to determine a the equation of a line.

Let's say the line passes through the points (2,3) and (5,7). The resulting system of linear equations, obtained by substituting the values in for x and y is shown. Simply solve the system of linear equations, plug in the values for A and B, and you have the model.

$$3 = 2A + B$$

$$7 = 5A + B$$

Solving that gives $A = 4/3$ and $B = 1/3$. The equation of the line passing through the given points is $y = 4/3\ x + 1/3$.

Quadratic Model: $y=Ax^2 + Bx + C$

This time, there are three variables, A, B, and C. Therefore, it takes three non-collinear points to determine a equation of a parabola.

Let's say the parabola passes through the points (2,3), (5,7), and (8,4). Substitute the values in for x and y and the three linear equations that result is the system that needs to be solved to find A, B, and C.

$$3 = 4A + 2B + C$$

$$7 = 25A + 5B + C$$

$$4 = 64A + 8B + C$$

If you're wondering where there 4, 25, and 64 came from, they are the values of x^2 when $x = 2$, $x = 5$, and $x = 8$.

Solving the system gives A = -7/18, B = 73/18, and C = -32/9. The equation is $y = -7/18\ x^2 + 73/18\ x - 32/9$. This procedure could be extended to find the equation of a cubic passing through four points or a quartic equation passing through five points.

Circle: $x^2 + y^2 + Dx + Ey + F = 0$

You have three variables, D, E, and F. Therefore, it takes three non-collinear points to determine the equation of a circle.

Let's say the circle passes through the points (2,3), (5,7), and (8,4). Substitute the values in for x and y and the equations that result give you the system of linear equations that can give the coefficients that define the circle. You may find it easier to move the constant to the other side.

$4 + 9 + 2D + 3E + F = 0$ becomes $2D + 3E + F = -13$

$25 + 49 + 5D + 7E + F = 0$ becomes $5D + 7E + F = -74$

$64 + 16 + 8D + 4E + F = 0$ becomes $8D + 4E + F = -80$

If you solve that system, you get D = -69/7, E = -55/7, and F = 212/7.

Substituting into the original model gives

$$x^2 + y^2 -69/7\ x - 55/7\ y + 212/7 = 0.$$

If you were so inclined (this is a pretty ugly problem), you could complete the square to find the centre and radius of the circle.

Least Squares Regression Parabola

If you have exactly three non-collinear points, then the parabola will pass through the points exactly. If there are more than three points, then you must fit a quadratic model to the data, but it won't necessarily pass through all of the points. Recall the least squares regression line that we discussed in the last section that went with the model $y = ax + b$.

$b\Sigma 1 + a\Sigma x = -y$

$b\Sigma x + a\Sigma x^2 = -xy$

Now, we are going to find a system of equations that will fit the model $y = ax^2 + bx + c$. It is an extension of the previous model. Notice it has the same pattern as before. As you work from left to right on the left hand side, each summation has one additional x in it. As you work from top to bottom, each term has one more x than the corresponding term in the previous equation.

$c\Sigma 1 + b\Sigma x + a\Sigma x^2 = -y$

$c\Sigma x + b\Sigma x^2 + a\Sigma x^3 = -xy$

$c\Sigma x^2 + b\Sigma x^3 + a\Sigma x^4 = -x^2y$

If you wanted to find a least squares regression cubic, then you would need four variables and at least five points (four points would exactly fit the model). Follow the same pattern as before, adding one more variable (for d) and one more row.

Partial Fractions

Addition of Rational Expressions

In arithmetic, you learned how to add fractions. You found the least common denominator, and then multiplied both the numerator and denominator of each term by what was needed to complete the common denominator.

$$\frac{1}{x-1}+\frac{2x-1}{x^2+3}=\left(\frac{1}{x-1}\right)\left(\frac{x^2+3}{x^2+3}\right)+\left(\frac{2x-1}{x^3+3}\right)\left(\frac{x-1}{x-1}\right)$$

$$=\frac{x^2+3+2x^2-3x+1}{(x-1)\left(x^3+3\right)}$$

$$=\frac{3x^2-3x+4}{(x-1)\left(x^2+3\right)}$$

In algebra, you have carried that process on to addition of rational expressions. You once again multiplied the numerator and denominator of each term by what was missing from the denominator of that term. With Partial Fraction Decomposition, we're going to reverse the process and decompose a rational expression into two or more simpler proper rational expressions that were added together.

Partial Fraction Decomposition

Partial Fraction Decomposition only works for proper rational expressions, that is, the degree of the numerator must be less than the degree of the denominator. If it is not, then you must perform long division first, and then perform the partial fraction decomposition on the rational part (the remainder over the divisor). After you've done the partial fraction decomposition, just add back in the quotient part from the long division. When discussing polynomials in section 3.4, we learned that every polynomial with real coefficients can be factored using only linear and irreducible quadratic factors. This means that there are only two types of factors that we have to worry about.

Linear Factors

If the partial fractions we're decomposing the rational expression into must be proper, then the only thing that can be over a linear factor is a constant. So, for every linear factor in the denominator, you will need a constant in over that in the numerator.

Irreducible Quadratic Factors.

If the partial fractions we're decomposing the rational expression into must be proper, then an irreducible quadratic factor could have a linear term and/or a constant term in the numerator. So, for every irreducible quadratic factor in the denominator, you will need a linear term and a constant term in the numerator.

Repeated Factors

Consider a fraction in which the denominator is 8. Does that mean that the denominator of each term being added together had to be an 8? No, the denominators could have been 2, 4, or 8 because the common denominator between 2, 4, and 8 is 8. The implications of this for partial fraction decomposition are that when you have a repeated factor (a factor with a multiplicity other than one), you need to include a factor in the expansion for each power possible.

For example, if you have an $(x-2)^3$, you will need to include an $(x-2)$, an $(x-2)^2$, and an $(x-2)^3$.

The exponents of 2 or 3 does not change whether the factor is linear or quadratic, only how many times the factor is there. Each of those $(x-2)$ factors would receive a constant term in the numerator because $x-2$ is linear, no matter what power it is raised to.

Improper Fractions

A proper fraction is a fraction where the numerator is less than the denominator. For rational expressions, that means the degree in the numerator is less than the degree in the denominator. If you have an improper fraction, you first have to perform long division to get a quotient and a remainder. Leave the quotient part alone, but perform the partial fraction decomposition on the fraction that remained.

The Basic Equation

After setting up the partial fraction equation, you multiply both sides of the equation by the least common denominator to get rid of the fractions. The resulting equation, without the fractions, is called the Basic Equation.

Technique 1 - Pick Values for x

$$\frac{x+2}{x(x-4)} = \frac{A}{x} + \frac{B}{x-4}$$

When there are linear factors, the easiest way to do the decomposition is probably to pick nice values for x.

In the example to the right, we take each factor in the denominator and give it its own term on the right side. Since each factor in the denominator is linear and the rational expressions need to be proper, a constant term was placed in each numerator of each term.

We multiply through by the least common denominator to arrive at the equation without the fractions. This equation is called the Basic Equation and is...

$$x + 2 = A (x - 4) + B x$$

"Nice" values for x are those that cause each linear factor to be zero. In this case, x=4 and x=0 are nice. "Nice" values will cause every term except for one to drop out of the equation, and so you will be able to find the value of a variable very quickly.

When you let x = 0, the B term drops out and you get...

0 + 2 = A (0 - 4) or 2 = -4A. Solving that gives A = -1/2.

When you let x = 4, the A term drops out and you get...

4 + 2 = B (4) or 6 = 4B. Solving that gives B = 3/2.

$$\frac{x+2}{x(x-4)} = \frac{-1/2}{x} + \frac{3/2}{x-4}$$

$$= \frac{1}{2}\left(\frac{3}{x-4} - \frac{1}{x}\right)$$

After you have found the values of each constant, it is important that you plug those values back into the decomposition. Don't just stop with A=-1/2, B=3/2 because someone else may have defined A and B differently. The correct answer is to place them back into the decomposition and simplify if necessary. The simplification would be to reduce the number of signs (don't have the first term negative if possible) and to eliminate the compound fractions by factoring out the least common denominator of the numerators.

Note: There are only as many nice values as there are distinct (different) linear factors. If there are repeated linear factors or irreducible quadratic factors (repeated or not), you will not have enough "nice" values to pick. In cases like that, you will have to pick convenient but not-so-nice values and then substitute the known constants into the equation to find the other constants. Plug in easy numbers like x=0, x=1, etc.

You will need to pick as many values for x as there are constants to be found.

Technique 2 - Creating a System of Linear Equations

The first technique of picking values for x works extremely well when all the factors are distinct linear factors. If there are any linear factors, then first method is still probably the better technique to use. However, if there are only irreducible quadratic factors, then the method of picking values for x can become messy.

$$\frac{2x^2+x+8}{(x^2+4)^2}=\frac{Ax+B}{x^2+4}+\frac{Cx+D}{(x^2+4)^2}$$

There is another way to do these problems (actually, this technique will work when there are linear factors, just that the other is easier and faster).

The first part of the process is the same. Go ahead and write out the decomposition, including constant terms over linear factors and linear and constant terms over the irreducible quadratic factors. Then, multiply through by the least common denominator to find the basic equation. That much is the same. The basic equation is...

$2x^2 + x + 8 = Ax (x^2 + 4) + B (x^2+4) + Cx + D$

Now the difference comes.

Go ahead and expand (multiply out) the basic equation...

$2x^2 + x + 8 = Ax^3 + 4Ax + Bx^2+ 4B + Cx + D$

and regroup the terms by common powers of the variable x.

$2x^2 + x + 8 = Ax^3 + Bx^2 + 4Ax + Cx + 4B + D$

Now factor out the x's by power.

$2x^2 + x + 8 = (A)\ x^3 + (B)\ x^2 + (4A + C)\ x + (4B + D)$ (1)

The next part works because if two polynomials are going to be equal, they must have the same number of similar terms on both sides. So the trick here is to equate both sides of the equation together by equating similar terms.

Set the x^3 terms on the left (of which there are none) equal to the x^3 terms on the right (of which there are A) to arrive at the first equation 0=A. Well, that was pretty easy, and already you have the value for A.

Set the x^2 terms on the left (of which there are 2) equal to the x^2 terms on the right (of which there are B) to arrive at the second equation of 2=B. Now you know B.

For the x's, there is 1 on the left and 4A+C on the right, so 1=4A+C.

For the constants, there are 8 on the left and 4B+D on the right, so 8=4B+D.

You may have to solve a system of linear equations, but that's why this section on partial fractions is in this chapter. We just refreshed our memories on how to solve a system of linear equations in the last section.

$$\frac{2x^2+x+8}{(x^2+4)^2}=\frac{0x+2}{x^2+4}+\frac{1x+0}{(x^2+4)^2}$$

$$=\frac{2}{x^2+4}+\frac{x}{(x^2+4)^2}$$

After some solving, we arrive at A=0, B=2, C=1, and D=0 and we are ready to plug those values back into the original decomposition to arrive at the final answer.

Be sure and simplify the answer if necessary.

This example was actually pretty easy because you were able to find A and B right off the bat. Many times, you will need to solve a much larger system of equations to find the values.

Systems of Inequalities

Graph of an Inequality

When we were dealing with one variable, the graph of an inequality was some interval on the real number line. Now we are dealing with two variables, and the graph of an inequality is the set of all ordered pairs (x, y) that satisfy the inequality. Typically, this will be represented by a half-plane.

Sketching an Inequality in Two Variables

1. Replace the inequality with an equal sign and sketch the resulting equation. Use a solid line if the equal to is included (<= or >=) and a dashed line of the inequality is strict (< or >).

2. Pick a test point not on the graph of the equation. If that test point satisfies the inequality, then shade every point on the side of the inequality that includes your test point. If the test point doesn't satisfy the inequality, then shade the other side of the inequality.

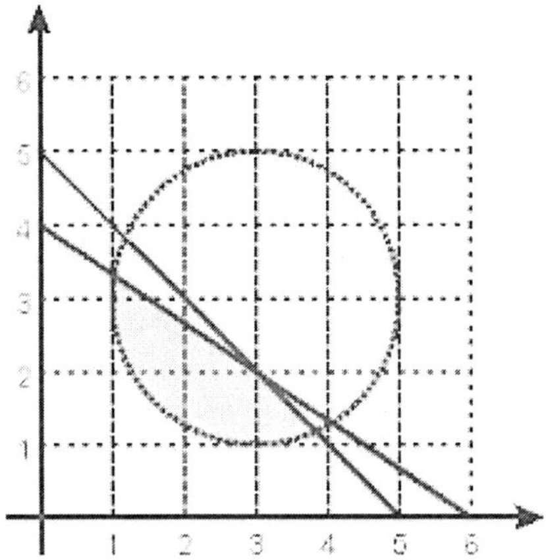

Figure: *Solving a System of Inequalities*

The solution to a system of inequalities is the set of all ordered pairs (x, y) that satisfy all the inequalities simultaneously.

The solution to the system of linear inequalities given is sketched to the right.

$$(x - 3)^2 + (y - 3)^2 < 4$$

$$x + y <= 5$$

$$2x + 3y <= 12$$

Notice the circle is dotted (dashed) instead of solid. This is because there is a strict inequality (< or >), that is, the equal to is not included.

Now, one misleading thing about generating the output on a computer is that it looks very easy to sketch the system. In reality, there are four ways you can sketch the solution to a system of linear inequalities

1. Draw all the lines and curves. Pick a test point in each region (there will be many regions - there are 8 in the figure shown). Drawback - way too much work.
2. Shade each inequality as it is encountered. Whatever is shaded for each inequality is the solution. If there are three inequalities, whatever is shaded three times is the solution. If there are five inequalities, then whatever is shaded five times is the solution.

Drawback - if you do a good job of shading, and you should, it becomes difficult to tell what is shaded three times as opposed to what is shaded four times.

3. Shade each inequality as it is encountered, but shade the false part in each case. The contrapositive of "if doesn't work, then shaded" is "if not shaded, then works". In other words, whatever is left unshaded will be the part that is the solution. Drawback - your answers will be unshaded where the books will be shaded.
4. Shade just the correct edge of each inequality. Then you can extend those to see which regions are shaded on all sides.

Whichever technique you use, right the word "yes" or something similar in the solution region.

Determining the System from the Graph.

1. Write the equation for each boundary on the graph.
2. Pick a test point in the solution region and substitute it into the equation.
3. Replace the equal sign by whichever inequality is needed to make the statement true. Include the equal sign if it is a solid line and don't include the equal sign in the inequality if it is a dashed line.

Linear Programming

In business, it is often desirable to find the production levels that will produce the maximum profit or the minimum cost. The production process can often be described with a set of linear inequalities called constraints. The profit or cost function to be maximized or minimized is called the objective function. The process of finding the optimal levels with the system of linear inequalities is called linear programming (as opposed to non-linear programming).

Definitions

Objective Function: The linear function (equal sign) representing cost, profit, or some other quantity to be maximized of minimized subject to the constraints.

Constraints: A system of linear inequalities.

Problem Constraints: The linear inequalities that are derived from the application. For example, there may be only 40 hours a machine can be used in a week, so the total time it is used would have to be <= 40. The problem constraints are usually stated in the story problem.

Non-Negativity Constraints

The linear inequalities x>=0 and y>=0.

These are included because x and y are usually the number of items produced and you cannot produce a negative number of items, the smallest number of items you could produce is zero.

These are not (usually) stated, they are implied.

Feasible Region

The solution to the system of linear inequalities. That is, the set of all points that satisfy all the constraints. Only points in the feasible region can be used.

Corner Point

A vertex of the feasible region. Not every intersection of lines is a corner point. The corner points only occur at a vertex of the feasible region. If there is going to be an optimal solution to a linear programming problem, it will occur at one or more corner points, or on a line segment between two corner points.

Bounded Region

A feasible region that can be enclosed in a circle. A bounded region will have both a maximum and minimum values.

Unbounded Region

A feasible region that can not be enclosed in a circle.

Fundamental Theorem of Linear Programming: Recall that almost every area of mathematics has its fundamental theorem.

Here are some of the fundamental theorems or principles that occur in your text.

Fundamental Theorem of Arithmetic: Every integer greater than one is either prime or can be expressed as an unique product of prime numbers.

Fundamental Theorem of Algebra : Every polynomial in one variable of degree n > 0 has at least one real or complex zero.

Fundamental Counting Principle : If there are m ways to do one thing, and n ways to do another, then there are m*n ways of doing both.

Fundamental Theorem of Linear Programming

If there is a solution to a linear programming problem, then it will occur at a corner point, or on a line segment between two corner points.

The Fundamental Theorem of Linear Programming is a great help. Instead of testing all of the infinite number of points in the feasible region, you only have to test the corner points. Whichever corner point yields the largest value for the objective function is the maximum and whichever corner point yields the smallest value for the objective function is the minimum.

Solving a Linear Programming Problem

If the problem is not a story problem, skip to step 3.

1. Define the variables. Usually, a good choice for the definition is the quantity they asked you to find in the problem.
2. Write the problem by defining the objective function and the system of linear inequalities. Don't forget about the non-negativity constraints where necessary.
3. Sketch the system of linear inequalities to obtain the feasible region.
4. Identify each corner point of the feasible region. You can find the corner points by forming a 2x2 system of linear equations from the two lines that intersect at that point and solving that system.
5. Evaluate the objective function at each corner point.
6. Choose the point yielding the largest value or smallest value depending on whether the problem is a maximization or minimization problem.

Be careful how you give the answer. The answer should give not only the maximum or minimum value (the value of the objective function), but it should also give the location where that extremum occurs. Example: The maximum value is 9 when x=2 and y=3. If it is a story problem, then give the answer in terms of the original definitions of x and y.

If the slope of the objective function is negative and you take a line with that slope passing through the origin and move it to the right through the feasible region, the last corner point hit by that moving line will be the maximum value.

In the example shown, the last line with slope m=-4/3 that touches the feasible region touches at the corner point (6,3).

Since z=4(6)+3(3)=24+9=33, the maximum value is 33 when x=6 and y=3.

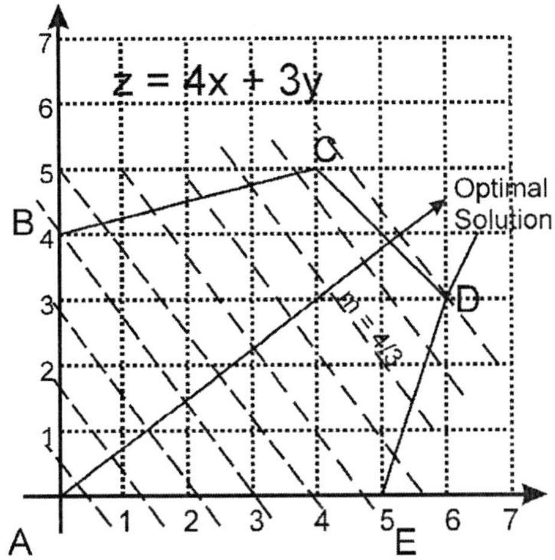

Figure: *Geometric Approach*

Algebraic Approach

Now, to verify the solution non-geometrically. Since we know the optimal solution has to occur at one or more corner points, we make a table listing all the corner points and evaluate the objective function at those points.

Corner Point	*x*	*y*	*z = 4x + 3y*
A	0	0	0
B	0	4	12
C	4	5	31
D	6	3	33
E	5	0	20

As you can see, the corner point with the maximum value is at (6,3).

We can also determine the minimum value from that table. A suitable answer, assuming the problem had asked for both the maximum and minimum is...

The minimum value is 0 when x=0 and y=0.

The maximum value is 33 when x=6 and y=3.

Chapter 2

Line Geometry

The notion of line or straight line was introduced by ancient mathematicians to represent straight objects with negligible width and depth. Lines are an idealization of such objects. Until the seventeenth century, lines were defined like this: "The line is the first species of quantity, which has only one dimension, namely length, without any width nor depth, and is nothing else than the flow or run of the point which [...] will leave from its imaginary moving some vestige in length, exempt of any width. [...] The straight line is that which is equally extended between its points."

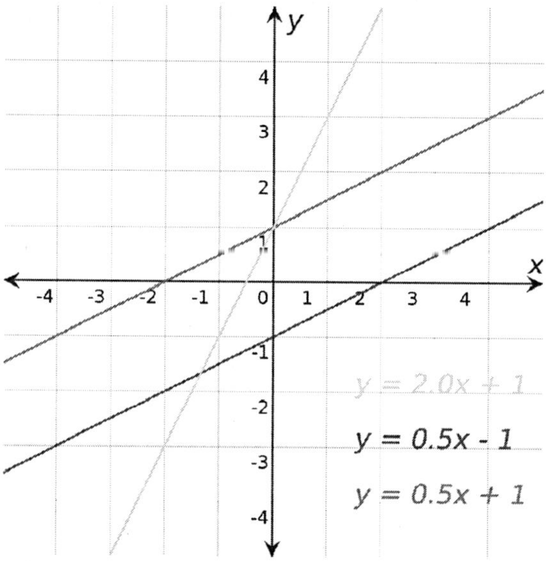

Figure: *Three lines — the red and blue lines have the same slope, while the red and green ones have same y-intercept.*

Euclid described a line as "breadthless length", and introduced several postulates as basic unprovable properties from which he constructed the geometry, which is now called Euclidean geometry to avoid confusion with other geometries which have been introduced since the end of nineteenth century (such as non-Euclidean geometry, projective geometry, and affine geometry).

In modern mathematics, given the multitude of geometries, the concept of a line is closely tied to the way the geometry is described. For instance, in analytic geometry, a line in the plane is often defined as the set of points whose coordinates satisfy a given linear equation, but in a more abstract setting, such as incidence geometry, a line may be an independent object, distinct from the set of points which lie on it.

When a geometry is described by a set of axioms, the notion of a line is usually left undefined (a so-called primitive object). The properties of lines are then determined by the axioms which refer to them. One advantage to this approach is the flexibility it gives to users of the geometry. Thus in differential geometry a line may be interpreted as a geodesic (shortest path between points), while in some projective geometries a line is a 2-dimensional vector space (all linear combinations of two independent vectors). This flexibility also extends beyond mathematics and, for example, permits physicists to think of the path of a light ray as being a line.

A line segment is a part of a line that is bounded by two distinct end points and contains every point on the line between its end points. Depending on how the line segment is defined, either of the two end points may or may not be part of the line segment. Two or more line segments may have some of the same relationships as lines, such as being parallel, intersecting, or skew.

Definitions versus Descriptions

All definitions are ultimately circular in nature since they depend on concepts which must themselves have definitions, a dependence which can not be continued indefinitely without returning to the starting point. To avoid this vicious circle certain concepts must be taken as primitive concepts; terms which are given no definition. In geometry, it is frequently the case that the concept of line is taken as a primitive. In those situations where a line is a defined concept, as in coordinate geometry, some other fundamental ideas are taken as primitives. When the line concept is a primitive, the behaviour and properties of lines are dictated by the axioms which they must satisfy.

In a non-axiomatic or simplified axiomatic treatment of geometry, the concept of a primitive notion may be too abstract to be dealt with. In this circumstance it is possible that a description or mental image of a primitive notion is provided to give a foundation to build the notion on which would formally be based on the (unstated) axioms. Descriptions of this type may be referred to, by some authors, as definitions in this informal style of presentation. These are not true definitions and could not be used in formal proofs of statements. The "definition" of line in Euclid's Elements falls into this category. Even in the case where a specific geometry is being considered (for example, Euclidean geometry), there is no generally accepted agreement among authors as to what an informal description of a line should be when the subject is not being treated formally.

Euclidean Geometry

When geometry was first formalised by Euclid in *Elements*, he defined lines to be "breadthless length" with a straight line being a line "which lies evenly with the points on itself". These definitions serve little purpose since they use terms which are not, themselves, defined. In fact, Euclid did not use these definitions in work and probably included them just to make it clear to the reader what was being discussed. In modern geometry, a line is simply taken as an undefined object with properties given by postulates.

In an axiomatic formulation of Euclidean geometry, such as that of Hilbert (Euclid's original axioms contained various flaws which have been corrected by modern mathematicians), a line is stated to have certain properties which relate it to other lines and points. For example, for any two distinct points, there is a unique line containing them, and any two distinct lines intersect at most one point. In two dimensions, ie. the Euclidean plane, two lines which do not intersect are called parallel. In higher dimensions, two lines that do not intersect may be parallel if they are contained in a plane, or skew if they are not.

Any collection of lines partitions the plane into convex polygons; this partition is known as an arrangement of lines.

Ray

If you define the concept of "order" of points of a line, you can define a *ray*, or half-line. A ray is part of a line which is finite in one direction, but infinite in the other. It can be defined by two points, the initial point, A, and one other, B. The ray is all the points in the line segment between A and B together with all points, C, on the line through A and B such that the points appear on the line in the order A, B, C.

In topology, a ray in a space X is a continuous embedding $R^+ \to X$. It is used to define the important concept of end of the space.

Coordinate Geometry

In coordinate geometry, lines in a Cartesian plane can be described algebraically by linear equations and linear functions. In two dimensions, the characteristic equation is often given by the *slope-intercept form*:

$$y = mx + c$$

where:

m is the slope of the line.

b is the y-intercept of the line.

x is the independent variable of the function y.

In three dimensions, a line is described by parametric equations:

$$x = x_0 + at$$

$$y = y_0 + bt$$

$$z = z_0 + ct$$

where:

x, y, and z are all functions of the independent variable t.

x_0, y_0, and z_0 are the initial values of each respective variable.

a, b, and c are related to the slope of the line, such that the vector (a, b, c) is a parallel to the line.

In R^2, every line L is described by a linear equation of the form

$$L = \{(x, y) \mid ax + by = c\}$$

with fixed real coefficients a, b and c such that a and b are not both zero. Important properties of these lines are their slope, x-intercept and y-intercept. The eccentricity of a straight line is infinity.

Euclidean Space

In Euclidean space R^n (and analogously in all other vector spaces), we define a line L as a subset of the form

$$L = \{\mathbf{a} + t\mathbf{b} \mid t \in \mathbb{R}\}$$

where a and b are given vectors in R^n with b non-zero. The vector b describes the direction of the line, and a is a point on the line. Different choices of a and b can yield the same line.

Projective Geometry

In projective geometry, a line is similar to that in Euclidean geometry but has slightly different properties. In many models of projective geometry, the idea of the line rarely conforms to the notion of the "straight curve" as it is visualised in Euclidean geometry. The Poincaré half-plane model is a typical example of this.

Geodesics

The "straightness" of a line, interpreted as the property that it minimizes distances between its points, can be generalized and leads to the concept of geodesics on differentiable manifolds.

Plane (Geometry)

In mathematics, a plane is a flat surface Chyea. Planes can arise as subspaces of some higher dimensional space, as with the walls of a room, or they may enjoy an independent existence in their own right, as in the setting of Euclidean geometry.

Figure: *Two intersecting planes in three-dimensional space*

When working in two-dimensional Euclidean space, the definite article is used, the plane, to refer to the whole space. Many fundamental tasks in geometry, trigonometry, and graphing are performed in two-dimensional space, or in other words, in the plane. A lot of mathematics can be and has been performed in the plane, notably in the areas of geometry, trigonometry, graph theory and graphing. All two-dimensional figures are assumed to be on a plane, even on *the* plane, unless otherwise specified.

Euclidean Geometry

Euclid set forth the first known axiomatic treatment of geometry. He selected a small core of undefined terms (called *common notions*) and postulates (or axioms) which he then uses these to prove various geometrical statements. Although the plane in its modern sense is not directly given a definition anywhere in the *Elements*, it may be thought of as part of the common notions. In his work Euclid never makes use of numbers to measure length, angle, or area. In this way the Euclidean plane is not quite the same as the Cartesian plane.

Planes in Various Areas of Mathematics

In addition to its familiar geometric structure, with isomorphisms that are isometries with respect to the usual inner product, the plane may be viewed at various other levels of abstraction. Each level of abstraction corresponds to a specific category.

At one extreme, all geometrical and metric concepts may be dropped to leave the topological plane, which may be thought of as an idealized homotopically trivial infinite rubber sheet, which retains a notion of proximity, but has no distances. The topological plane has a concept of a linear path, but no concept of a straight line. The topological plane, or its equivalent the open disc, is the basic topological neighbourhood used to construct surfaces (or 2-manifolds) classified in low-dimensional topology. Isomorphisms of the topological plane are all continuous bijections. The topological plane is the natural context for the branch of graph theory that deals with planar graphs, and results such as the four colour theorem.

The plane may also be viewed as an affine space, whose isomorphisms are combinations of translations and non-singular linear maps. From this viewpoint there are no distances, but colinearity and ratios of distances on any line are preserved.

Differential geometry views a plane as a 2-dimensional real manifold, a topological plane which is provided with a differential structure. Again in this case, there is no notion of distance, but there is now a concept of smoothness of maps, for example a differentiable or smooth path (depending on the type of differential structure applied). The isomorphisms in this case are bijections with the chosen degree of differentiability.

In the opposite direction of abstraction, we may apply a compatible field structure to the geometric plane, giving rise to the complex plane and the major area of complex analysis. The complex field has only two isomorphisms that leave the real line fixed, the identity and conjugation. In the same way as in the real case, the plane may also be viewed as the simplest, one-dimensional (over the complex numbers) complex manifold, sometimes called the complex line. However, this viewpoint contrasts sharply with the case of the plane as a 2-dimensional real manifold. The isomorphisms are all conformal bijections of the complex plane, but the only possibilities are maps that correspond to the composition of a multiplication by a complex number and a translation.

In addition, the Euclidean geometry (which has zero curvature everywhere) is not the only geometry that the plane may have. The plane may be given a spherical geometry by using the stereographic projection. This can be thought of as placing a sphere on the plane (just like a ball on the floor), removing the top point, and projecting the sphere onto the plane from this point). This is one of the projections that may be used in making a flat map of part of the Earth's surface. The resulting geometry has constant positive curvature.

Alternatively, the plane can also be given a metric which gives it constant negative curvature giving the hyperbolic plane. The latter possibility finds an application in the theory of special relativity in the simplified case where there are two spatial dimensions and one time dimension. (The hyperbolic plane is a timelike hypersurface in three-dimensional Minkowski space.)

Topological and Differential Geometric Notions

The one-point compactification of the plane is homeomorphic to a sphere; the open disk is homeomorphic to a sphere with the "north pole" missing; adding that point completes the (compact) sphere. The result of this compactification is a manifold referred as the Riemann sphere or the complex projective line. The projection from the Euclidean plane to a sphere without a point is a diffeomorphism and even a conformal map.

The plane itself is homeomorphic (and diffeomorphic) to an open disk. For the Lobachevsky plane such diffeomorphism is conformal, but for the Euclidean plane it is not.

Planes in Fiction

The 1884 novel Flatland by Edwin A. Abbott features the concept of a geometric, two-dimensional infinite plane inhabited by living geometric figures (triangles, squares, circles, etc.).

It has been described by Isaac Asimov, in his foreword to the Signet Classics 1984 edition, as "the best introduction one can find into the manner of perceiving dimensions." this is not a actual plane in the mathematics book.

Geometry

The word *geometry* is Greek for *geos* - meaning earth and *metron* - meaning measure. Geometry was extremely important to ancient societies and was used for surveying, astronomy, navigation, and building. Geometry, as we know it is actually known as Euclidean geometry which was written well over 2000 years ago in Ancient Greece by Euclid, Pythagoras, Thales, Plato and Aristotle just to mention a few. The most fascinating and accurate geometry text was written by Euclid, and was called Elements. Euclid's text has been used for over 2000 years!

Geometry is the study of angles and triangles, perimeter, area and volume. It differs from algebra in that one develops a logical structure where mathematical relationships are proved and applied. In part 1, you will learn about the basic terms associated with Geometry.

Some Basics and Important Postulates

Background: Euclid of Alexandria wrote 13 books called 'The Elements' around 300 BC.

These books laid the foundation of geometry. Some of the postulates below were actually posed by Euclid in his 13 books.

They were assumed as axioms, without proof. Euclid's postulates have been slightly corrected over a period of time. Some are listed here and continue to be part of 'Euclidean Geometry'.

Know this stuff! Learn it, memorize it and keep this page as a handy reference if you expect to understand Geometry.

There are some basic facts, information and postulates that are very important to know in geometry. Not everything is proved in Geometry, thus we use some *postulates* which are basic assumptions or unproved general statements that we accept. Here are a few of the basics and postulates that are intended for entry level Geometry.

Conic Section

In mathematics, a conic section (or just conic) is a curve obtained by intersecting a cone (more precisely, a circular conical surface) with a plane. A conic section is therefore a restriction of a quadric surface to the plane. The conic sections were named and studied as long ago as 200 BC, when Apollonius of Perga undertook a systematic study of their properties.

Eccentricity

The four defining conditions above can be combined into one condition that depends on a fixed point F (the *focus*), a line L (the *directrix*) not containing F and a nonnegative real number e (the *eccentricity*). The corresponding conic section consists of all points whose distance to F equals e times their distance to L. For $0 < e < 1$ we obtain an ellipse, for $e = 1$ a parabola, and for $e > 1$ a hyperbola.

For an ellipse and a hyperbola, two focus-directrix combinations can be taken, each giving the same full ellipse or hyperbola. The distance from the centre to the directrix is *a*/*e*, where is the semi-major axis of the ellipse, or the distance from the centre to the tops of the hyperbola. The distance from the centre to a focus is. In the case of a circle, the eccentricity $e = 0$, and one can imagine the directrix to be infinitely far removed from the centre. However, the statement that the circle consists of all points whose distance is e times the distance to L is not useful, because we get zero times infinity.

The eccentricity of a conic section is thus a measure of how far it deviates from being circular. For a given, the closer is to 1, the smaller is the semi-minor axis.

Cartesian Coordinates

In the Cartesian coordinate system, the graph of a quadratic equation in two variables is always a conic section, and all conic sections arise in this way. The equation will be of the form

$Ax^2 + Bxy + Cy^2 + Dx + Ey + F = 0$ with A, B, C not all zero.

then:

- if $B^2 - 4AC < 0$, the equation represents an ellipse (unless the conic is degenerate, for example $x^2 + y^2 + 10 = 0$);
  - o if $A = C$ and $B = 0$, the equation represents a circle;
- if $B^2 - 4AC = 0$, the equation represents a parabola;
- if $B^2 - 4AC > 0$, the equation represents a hyperbola;

o if we also have $A + C = 0$, the equation represents a rectangular hyperbola.

Note that A and B are just polynomial coefficients, not the lengths of semi-major/minor axis as defined in the following sections.

In matrix notation the equation above becomes:

$$[x \quad y].\begin{bmatrix} A & B/2 \\ B/2 & C \end{bmatrix}.\begin{bmatrix} x \\ y \end{bmatrix} + Dx + Ey + F = 0.$$

or

$$[x \quad y \quad 1].\begin{bmatrix} A & B/2 & D/2 \\ B/2 & C & E/2 \\ D/2 & E/2 & F \end{bmatrix}.\begin{bmatrix} x \\ y \\ 1 \end{bmatrix} = 0.$$

and

$$B^2 - 4AC = -4\begin{vmatrix} A & B/2 \\ B/2 & C \end{vmatrix}.$$

Through change of coordinates these equations can be put in standard forms:

- Circle: $x^2 + y^2 = a^2$
- Ellipse: $\frac{x^2}{a^2} + \frac{y^2}{b^2} = 1$
- Parabola: $y^2 = 4ax, x^2 = 4ay$,
- Hyperbola: $\frac{x^2}{a^2} - \frac{y^2}{b^2} = 1, \frac{x^2}{b^2} - \frac{y^2}{a^2} = -1$
- Rectangular Hyperbola: $xy = c^2$

Such forms will be symmetrical about the x-axis and for the circle, ellipse and hyperbola symmetrical about the y-axis.

The rectangular hyperbola however is only symmetrical about the lines $y = x$ and $y = -x$. Therefore its inverse function is exactly the same as its original function.

These standard forms can be written as parametric equations,

- Circle: $(a\cos\theta, a\sin\theta)$,
- Ellipse: $(a\cos\theta, b\sin\theta)$,
- Parabola: $(at^2, 2at)$,
- Hyperbola: $(a\sec\theta, b\tan\theta)$ or $(\pm a\cosh u, b\sinh u)$.
- Rectangular Hyperbola: $\left(ct, \frac{c}{t}\right)$

Homogeneous Coordinates

In homogeneous coordinates a conic section can be represented as:

$$A_1x^2 + A_2y^2 + A_3z^2 + 2B_1xy + 2B_2xz + 2B_3yz = 0.$$

Or in matrix notation

$$[x \quad y \quad z].\begin{bmatrix} A_1 & B_1 & B_2 \\ B_1 & A_2 & B_3 \\ B_2 & B_3 & A_3 \end{bmatrix}.\begin{bmatrix} x \\ y \\ z \end{bmatrix} = 0.$$

The matrix $M = \begin{bmatrix} A_1 & B_1 & B_2 \\ B_1 & A_2 & B_3 \\ B_2 & B_3 & A_3 \end{bmatrix}$ is called *the matrix of the conic section.*

$\Delta = \det(M) = \det\left(\begin{bmatrix} A_1 & B_1 & B_2 \\ B_1 & A_2 & B_3 \\ B_2 & B_3 & A_3 \end{bmatrix}\right)$ is called the determinant of the conic section.

If Ä = 0 then the *conic section* is said to be *degenerate*, this means that the conic section is in fact a union of two straight lines. A conic section that intersects itself is always degenerate, however not all degenerate conic sections intersect themselves, if they do not they are straight lines.

For example, the conic section $[x \quad y \quad z].\begin{bmatrix} 1 & 0 & 0 \\ 0 & -1 & 0 \\ 0 & 0 & 0 \end{bmatrix}.\begin{bmatrix} x \\ y \\ z \end{bmatrix} = 0$ reduces to the union of two lines:

$$\{x^2 - y^2 = 0\} = \{(x+y)(x-y) = 0\} = \{x+y=0\} \cup \{x-y=0\}.$$

Similarly, a conic section sometimes reduces to a (single) line:

$$\{x^2 + 2xy + y^2 = 0\} = \{(x+y)^2 = 0\} = \{x+y=0\} \cup \{x+y=0\} = \{x+y=0\}..$$

$\delta = \det\left(\begin{bmatrix} A_1 & B_1 \\ B_1 & A_2 \end{bmatrix}\right)$ is called the discriminant of the conic section.

If δ = 0 then the *conic section* is a parabola, if δ<0, it is an hyperbola and if δ>0, it is an ellipse. A conic section is a circle if δ>0 and $A_1 = A_2$, it is an rectangular hyperbola if δ<0 and $A_1 = -A_2$.

It can be proven that in the complex projective plane CP^2 two conic sections have four points in common (if one accounts for multiplicity), so there are never more than 4 intersection points and there is always 1 *intersection point* (possibilities: 4 distinct intersection points, 2 singular intersection points and 1 double intersection points, 2 double intersection points, 1 singular intersection point and 1 with multiplicity 3, 1 intersection point with multiplicity 4).

If there exists at least one intersection point with multiplicity > 1, then the two conic sections are said to be tangent. If there is only one intersection point, which has multiplicity 4, the two conic sections are said to be osculating.

Furthermore each straight line intersects each conic section twice. If the intersection point is double, the line is said to be tangent and it is called the tangent line. Because every straight line intersects a conic section twice, each conic section has two points at infinity (the intersection points with the line at infinity). If these points are real, the conic section must be a hyperbola, if they are imaginary conjugated, the conic section must be an ellipse, if the conic section has one double point at infinity it is a parabola. If the points at infinity are (1,i,0) and (1,-i,0), the conic section is a circle. If a conic section has one real and one imaginary point at infinity or it has two imaginary points that are not conjugated it is neither a parabola nor an ellipse nor a hyperbola.

Chapter 3

Applications of Trigonometry

The Law of Sines

Given three non-collinear points, A, B and C, there is only one circle containing those three points. Three non-collinear points also form the vertices of a triangle. According to a theorem from geometry, the ratio of a side of a triangle to the sine of the opposite angle equals the diameter of the circumscribed circle. Since for a given triangle there is only one circumscribed circle, then one may compute its diameter by computing the ratio between *any* of its three sides and the sine of the angle opposite that side. That is,

$$d = a \,/\, \sin \mathrm{A} = b \,/\, \sin \mathrm{B} = c \,/\, \sin \mathrm{C}.$$

This identity is called the Law of Sines.

The Law of Sines, together with the Law of Cosines which we will study shortly, are useful for solving triangles. There are six pieces of information that must be known before a triangle is 'solved': the size of its three angles and the size of its three sides. In order to solve a triangle, one needs to have at least three of the six pieces of information to begin with. The one exception to this principle is this: knowing only the three angles is insufficient for finding the three sides

Why can't one determine the length of the three sides of a triangle if one knows all three angles?

Solution

There is such a thing as *similar triangles*: triangles of different sizes, yet which have congruent angles. For example, two equilateral triangles of different sizes still have all their angles equal to 60°. So just knowing the angles tells us nothing about the size of the sides.

There are six different ways to classify the three given pieces of information:

SSS: The lengths of all three sides are given.

SAS: The lengths of two sides and the size of the angle between them are given.

SSA: The lengths of two sides and the size of the angle opposite one of them are given.

ASA: The sizes of two angles and the length of the side between them are given.

SAA: The sizes of two angles and the length of the side opposite one of them are given.

AAA: The sizes of all three angles are given.

Note that some combinations of three pieces of information will not have a solution. For example, the sum of the three angles of a triangle must be 180°.

Thus in the case of ASA, if the two angles add to more than 180°, there will be no solution. Or in the case of SSS, if the sum of two of the sides is less than the third side, there will be no solution, since the sum of two sides of a triangle is always greater than the third side. SAS is the only case where all possible lengths of the sides and all possible angles greater than 0° and less than 180° will have a solution.

Exercise : Explain why each of the following 'triangles' has no solution:

(a) SSS: $a = 7$, $b = 2$, $c = 4$

(b) ASA: A = 85°, $b = 10$, C = 98°

(c) SAA: $a = 12$, B = 100°, C = 90°

(d) AAA: A = 65°, B = 82°, C = 35°

Solution

(a) The sum of two sides of a triangle must be larger than the third. Bud $b + c = 6$ and $a = 7$. So there is no solution.

(b) The sum of two angles of a triangle must be smaller than 180°. But A + B = 183°, so there is no solution.

(c) The sum of B and C is greater than 180° so there is no solution.

(d) The sum of the three angles is greater than 180°, so there is no solution.

In order to apply the Law of Sines, it is necessary to know the size of at least one side and the size of the angle opposite that side. That is, one must know at least one side-sine pair.

One will know a side-sine pair in the case of SSA or SAA.

But if one is given ASA, one can deduce the third angle since the sum all three angles must be 180°. Thus one will know a side-sine pair given ASA.

So the Law of Sines is used to solve triangles given SSA, SAA or ASA.

SAA and ASA are worked essentially the same: find the third angle, then use the side-sine pair to solve the other two sides.

Example: Given a = 10, A = 30° and C = 120°, solve the triangle.

First find B = 180° - 30° - 120° = 30°. Next, use the fact that b/sin B = a/sin A to find b, since b = a sin B/sin A = 10 sin (30°)/sin (30°) = 10.

Likewise, c/sin C = a/sin A, so c = a sin C/sin A = 10 sin (120°)/ sin (30°)

$=10\left(\sqrt{3}/2\right)\left(1/2\right)=5\sqrt{3}/2.$ Thus the solution of the triangle is:

B = 120°

b = 10

c = $5\sqrt{3}/2$

Exercise

Given A = 24°, B = 80° and c = 20, solve the triangle.

Solution

A = 24°, B = 80°, c = 20

At first glance it might appear that we do not have a side-sine pair.

But we can easily get the angle C opposite the given side and then apply the Law of Sines.

This problem is of the type SAA.

C = 180° - 24° - 80° = 76°

$$\frac{a}{\sin 24°}=\frac{20}{\sin 76°}$$

$$a=\frac{20\sin 24°}{\sin 76°}=8.38$$

$$\frac{b}{\sin 80^\circ} = \frac{20}{\sin 76^\circ}$$

$$b = \frac{20 \sin 80^\circ}{\sin 76^\circ} = 20.30$$

It is possible for the case SSA to have no solution for some combinations of lengths of the sides and angles.

Exercise

Explain why the following SSA 'triangle' has no solution. Hint: the sine of an angle cannot be greater than 1.

A = 60°, a = 2, b = 10

Solution

A = 60°, a = 2, b = 10

$$\frac{\sin B}{10} = \frac{\sin 60^\circ}{2} = \frac{\sqrt{3}}{4}$$

$$\sin B = \frac{5\sqrt{3}}{2} \approx 4.33 > 1$$

Since the sine of an angle cannot exceed 1, there can be no solution to this triangle.

Knowing the sine of one of the angles of a triangle is not sufficient information to uniquely determine the angle.

For example, suppose we want to find A and all we know is that sin A = 1/2. Then A could be 30° or A could be 150°.

We cannot assume that the correct value is 30° unless we can rule out 150°.

In general, if sin A = y, then either A = arcsin y or A = 180° - arcsin y. Sometimes this situation arises when solving SSA triangles.

When this happens and when we cannot rule out the larger solution, then we must list *two separate solutions* to the triangle.

Exercise

Find *two* solutions to the triangle:

a = 10, b = 4, B = 20°

Solution:

a = 10, b = 4, B = 20°

$$\frac{\sin A}{10} = \frac{\sin 20°}{4}$$

$$\sin A = \frac{10 \sin 20°}{4} \approx 0.8550$$

There are two angles between 0° and 180° whose sine is 0.8550: 58.8° and 121.2°.

Since the larger of the two angles, when added to the given angle 20° is only 141.2°, which is still less than 180°, the larger angle is a possible solution.

There will be two different solutions to this triangle.

Solution #1

A = 58.8°

C = 180° - 20° - 58.8° = 101.2°

$$\frac{c}{\sin 101.2°} = \frac{4}{\sin 20°}$$

$$c = \frac{4 \sin 101.2°}{\sin 20°} = 11.4$$

Solution #2

A = 121.2°

C = 180° - 20° - 121.2° = 38.8°

$$\frac{c}{\sin 38.8°} = \frac{4}{\sin 20°}$$

$$c = \frac{4 \sin 38.8°}{\sin 20°} = 7.3$$

Exercise

A ship sailing due north spots a lighthouse 30° to the left of its line of travel. Two miles later, the lighthouse is 50° to the left of its line of travel. How far is the ship from the lighthouse at that point? (Assume the earth is flat.)

Solution

$$\frac{a}{\sin 30°} = \frac{2}{\sin 20°}$$

$$a = \frac{2 \sin 30°}{\sin 20°} \approx 2.92 \text{ miles}$$

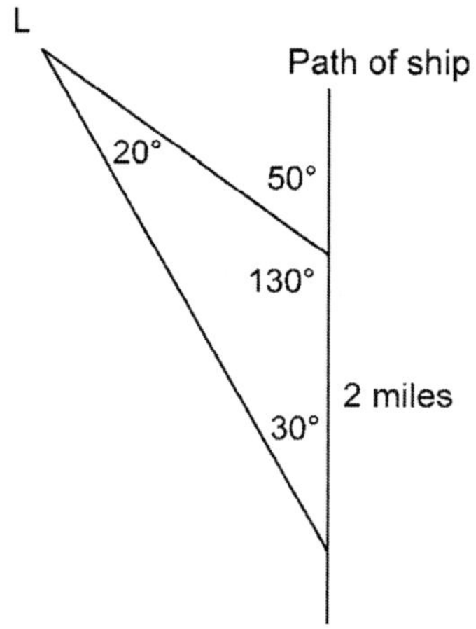

Exercise

Scientists wish to measure the diameter of a large circular meteor crater. Two points A and B on the edge of the pit are 120 feet apart. From a point C on the far side of the pit, the angle between the lines AC and BC is measured to be 8°. Use the Law of Sines to find the diameter of the crater.

Solution

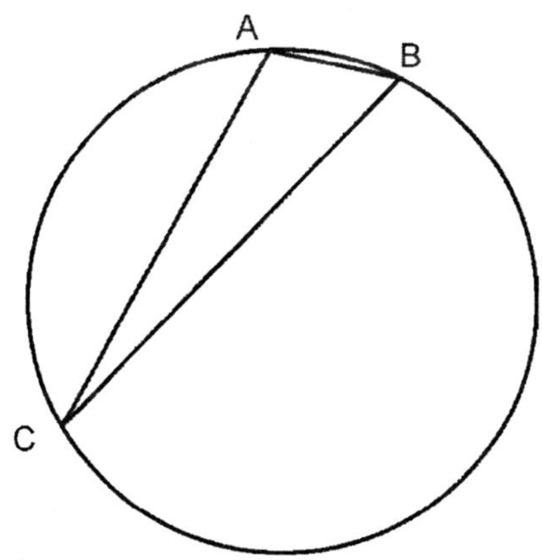

Remember, the Law of Sines states that the diameter of the circumscribed circle for a triangle ABC equals any side of the triangle divided by the sine of the opposite angle. Since we know the side AB has length 120 feet and the angle opposite the side AB is 8°, then the diameter must be

$$D = \frac{120}{\sin 80^\circ} = 862.2\text{feet}$$

The Law of Cosines

The Law of Cosines is a generalization of the Pythagorean Theorem. In a triangle, if C is a right angle, then $a^2 + b^2 = c^2$. The Law of cosines states that

$$a^2 + b^2 - 2\,ab\cos C = c^2.$$

But if C is not 90°, then that angle has lost its special Status. Thus the Law of Cosines must apply to other given variables.

So

$$b^2 + c^2 - 2\,bc\cos A = a^2 \text{ and}$$

$$a^2 + c^2 - 2\,ac\cos B = b^2.$$

The Law of Cosines may be used to solve triangles of type SAS and SSS.

To solve SAS problems, use the Law of Cosines to find the length of the side opposite the given angle.

Example:, given $a = 10$, B = 32° and $c = 15$.

The first step is to find the side opposite the given angle, in this case, b.

$$b^2 = a^2 + c^2 - 2\,ac\cos B. \text{ Thus,}$$

$$b^2 = 100 + 225 - 2\,(10)\,(15)\cos 32^\circ$$

$$= 325 - 300\,(0.848048096) = 325 - 254.4 = 70.59.$$

Thus $b = 8.40$, approximately.

Now, we must find the sizes of angles A and C. We can use the Law of Sines, *provided we find the angle opposite the smaller side first.* Why is it important to find the angle opposite the smaller of a and c first? If the triangle has an obtuse angle (i.e. > 90°), then it will be opposite the largest side of the triangle.

So we know that the angle opposite the smaller of the two given sides *cannot* be obtuse. Why should we be concerned whether an angle is obtuse or not?

Remember that knowing the sine of an angle of a triangle does not tell you what the angle is, since there are *two solutions* to an equation sin D = y between 0° and 180°, an acute angle D = arcsin y and an obtuse angle D = 180° - arcsin y. If we know in advance that the angle D is acute, then we know to use D = arcsin y.

In the example, we know $a = 10$, B = 32°, $c = 15$ and $b = 8.40$. Since a is smaller than c, we find A next using the Law of Sines.

sin A/10 = sin 32°/8.40

sin A = 0.6309

A = arcsin 0.6309 = 39.1°

Then the remaining angle can be found

C = 180° - 32° - 39.1° = 108.9°

So the solution is b = 8.40, A = 39.1°, C = 108.9°.

Exercise

Given A = 28°, b = 14, c = 10 solve the triangle.

Solution

A = 28°, b = 14, c = 10

Since we do not have a side-sine pair, we must use the Law of Cosines. This problem is of the SAS type: we are given two sides and the included angle (the angle between the two given sides).

First, we will solve for the side opposite the given angle :

$$a^2 = b^2 + c^2 - 2bc\cos A = 196 + 100 - 280\cos(28°) \approx 48.77$$

$$a = 6.98$$

Next we will use the Law of Sines to solve the angle opposite the *smaller* of the two given sides. It's important that we solve for the angle opposite the smaller of the two sides, since we know that it will not be obtuse. The angle opposite the largest side of a triangle may be an obtuse angle, that is, it may be greater than 90°. It is best to find obtuse angles last by subtracting the other two angles from 180°.

$$\frac{\sin C}{10} = \frac{\sin 28°}{6.98}$$

$$\sin C = \frac{10\sin 28°}{6.98} = 0.6722$$

$$C \approx 42.2°$$

B = 180° - 28° - 42.2° = 109.8°

Exercise

An engineer wishes to measure the diameter of a hole in the ground, but his tape measure isn't long enough. He places stakes at points A and B on opposite sides of the hole, then places a third stake at a point C at the edge of the hole somewhere between A and C. His tape is long enough to measure the distance from C to A at 28 feet and the distance from C to B at 26 feet. From the point C, the angle between the lines AC and CB is 158°. How far is it from A to B?

Solution

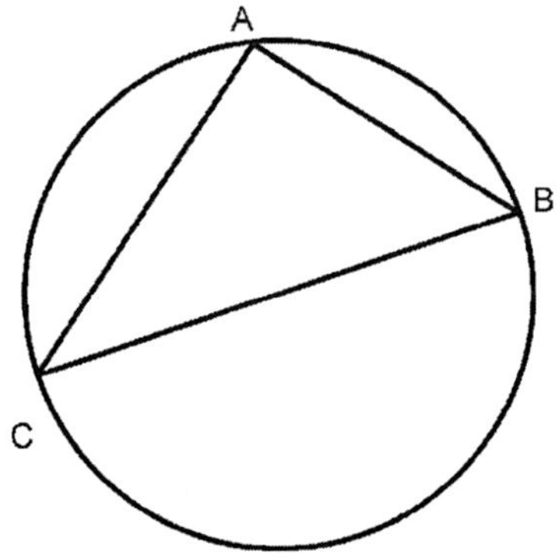

$| AB |^2 = 28^2 + 26^2 - 2(28)(26)\cos 158° = 2810.0$

So, | AB | H" 53 feet

There is an alternate form of the Law of Cosines which is used to solve the SSS case. In the standard form of the Law of Cosines, each of the three equations is solved for the cosine of the angle to get

$$\cos A = (b^2 + c^2 - a^2)/(2\ bc)$$

$$\cos B = (a^2 + c^2 - b^2)/(2\ ac)$$

$$\cos C = (a^2 + b^2 - c^2)/(2\ ab)$$

Thus, given the lengths of the three *sides* of a triangle, we can find the cosines of the three *angles.*

Knowing the cosine of an angle of a triangle, can we deduce the angle itself?

If we know that cos D = x, where D is an interior angle of a triangle, can we deduce that D = arccos x ?

Recall that the principle domain of the cosine function (for the purpose of finding the inverse function) is the interval [0, 180°].

This means that there is *only one* angle between 0° and 180° whose cosine is x. Thus, we can conclude that

$$A = \arccos[\ (b^2 + c^2 - a^2)/(2\ bc)\ ]$$

$$B = \arccos[\ (a^2 + c^2 - b^2)/(2\ ac)\ ]$$

$$C = \arccos[\ (a^2 + b^2 - c^2)/(2\ ab)\ ]$$

Typically, we use the Law of Cosines to find A and B, then subtract from 180° to find C.

Exercise

Given $a = 8$, $b = 5$ and $c = 4$, solve the triangle.

Solution

$$A = \arccos\left(\frac{b^2 + c^2 - a^2}{2bc}\right) = \arccos\left(-\frac{23}{40}\right) = 125.1°$$

$$B = \arccos\left(\frac{a^2 + c^2 - b^2}{2ac}\right) = \arccos\left(\frac{55}{64}\right) = 30.8$$

$$C = 180° - 125.1° - 30.8° = 24.1°$$

There is a formula from antiquity for computing the area of a triangle from the lengths of its three sides called Heron's Formula (also called Hero's Formula). I makes use of the idea of the *semiperimeter.* The semiperimeter is simply *half* the perimeter.

Given a triangle ABC with sides a, b and c and semiperimeter s, the area can be given as: Area = $[\ s\ (s - a)(s - b)(s - c)\ ]^{1/2}$

Exercise

Given a triangle ABC with sides $a = 8$, $b = 5$ and $c = 4$, find the area of the triangle.

Solution

$s = (1/2)\ (a + b + c) = 8.5$

Area = – $[\ s\ (s - a)(s - b)(s - c)\ ] = 9.045$

Trigonometry and Complex Numbers

Let $z = a + b\ i$ denote a complex number with real part a and imaginary part b. Then the absolute value of z, denoted $|\ z\ |$, is the distance from z to zero in the complex plane and is computed by the formula $|z| = \sqrt{a^2 + b^2}$

The line containing 0 and z makes an angle t with the real axis satisfying

$\tan\ (t) = b/a$. Furthermore,

$\cos\ (t) = a/|\ z\ |$ and

$\sin\ (t) = b/|\ z\ |$. Thus

$a = |\ z\ |\ \cos\ (t)$ and $b = |\ z\ |\ \sin\ (t)$, so

$z = |\ z\ |\ \cos\ (t) + |\ z\ |\ \sin\ (t)\ i$. Thus

$z = |\ z\ |\ [\ \cos\ (t) + i\ \sin\ (t)\ ]$

This is called the *trigonometric form* of a complex number.

Suppose $z1$ and $z2$ are complex numbers. Then their product is

$$z_1 . z_2 = | z_1 | [\cos (t_1) + i \sin (t_1)] | z_2 | [\cos (t_2) + i \sin (t_2)]$$
$$= | z_1 | | z_2 | \{[\cos (t_1) \cos (t_2) - \sin (t_1) \sin (t_2)]$$
$$+ i [\cos (t_1) \sin (t_2) + \sin (t_1) \cos (t_2)] \}$$
$$= | z_1 | | z_2 | [\cos (t_1 + t_2) + i \sin (t_1 + t_2)].$$ Thus,

$$z_1 . z_2 = | z_1 | | z_2 | [\cos (t_1 + t_2) + i \sin (t_1 + t_2)]$$

Exercise

Let $z = 3 + 3 i$. Find $| z |$. Find an angle t such that $z = | z | [\cos (t) + i \sin (t)]$. Write z in trigonometric form.

Solution

The magnitude of z is found by taking the square root of the sum of the squares if its components, thus:

$| z | = 3"2$

The angle is the standard angle with terminal side through the point (3, 3), or $\pi/4$. Thus, the trigonometric form for the complex number $3 + 3 i$ is:

$3"2 [\cos (\pi/4) + i \sin (\pi/4)]$

Exercise

Show that $z^2 = | z |^2 [\cos (2 t) + i \sin (2 t)]$.

Solution

$z = | z | (\cos t + i \sin t)$

So $z^2 = |z|^2 [\cos (t + t) + i \sin (t + t)] = | z |^2 [\cos (2 t) + i \sin (2 t)]$

The result in Exercise 6.3.2 can be extended to the general result, called DeMoivre's Theorem:

If $z = | z | [\cos (t) + i \sin (t)]$, then $z^n = | z |^n [\cos (n t) + i \sin (n t)]$.

Example: Find $(-1 + 2 i)^5$

Let $z = -1 + 2 i$. Then find z^5. First, we must find $|z| = \sqrt{1^2 + 2^2} = \sqrt{5}$ which is approximately 2.236067977. Next, we must find the angle t. Since z lies in the second quadrant of the complex plane, we are looking for an angle t in the second quadrant satisfying $\tan (t) = b/a = -2$. arctan (-2) will not do, since it is in quadrant IV. We want pi + arctan (-2), which is approximately 2.034443936. [Your calculator must be in radian mode.] Thus, $(-1 + 2\,i)^5$ is approximately equal to

$(2.236067977)^5$ [cos 10.17221968 + i sin 10.17221968]

= 55.90169937 [-0.733430296 - 0.679764665 i]

= - 40.99999992 - 37.99999999 i.

Since the real and imaginary parts of $(-1 + 2\,i)^5$ must be integers, the answer is (-1 + 2 i)5 = - 41 - 38 i.

Exercise

Find $(1 + i)^7$.

Solution

First we must put (1 + i) in trigonometric form. The magnitude is “2 and the angle is 45°. The seventh power of the complex number will have a magnitude which is the seventh power of “2, or 8”2. The angle of the seventh power will be seven times the angle 45°, or 315°. Thus, the seventh power of (1 + i) is

8"2 (cos 315° + i sin 315°) = 8 – 8 i

The inverse of finding powers of complex numbers is finding roots of complex numbers.

A complex number has two square roots, three cube roots, four fourth roots, etc. Generally, a complex number has n nth roots. All n of the nth roots of a complex number z are evenly spaced around a circle centred at 0 and having a radius equal to the nth root of the absolute value of z. Furthermore, if $z = |z|$ [cos (t) + i sin (t)], then one of the n nth roots of z is $|z|^{1/n}$ [cos (t/n) + i sin (t/n)]. Since the n nth roots are evenly spaced around a circle, they are separated by an angle of $2\pi / n$. So, once we find the root $|z|^{1/n}$ [cos (t/n) +i sin (t/n)], we find the other n - 1 roots by adding multiples of $2\pi / n$ to t/n.

Example: Find the cube roots of 1 + i.

First, write 1 + i in trigonometric form:

$$\sqrt{2}\left[\cos\left(\frac{\pi}{4}\right)+i\sin\left(\frac{\pi}{4}\right)\right]$$

Then one root is $\sqrt[6]{2}\left[\cos\left(\frac{\pi}{12}\right)+i\sin\left(\frac{\pi}{12}\right)\right]$. To find the two remaining cube roots, we add multiples of to to get and . Thus the three cube roots of 1 + i are

$$\sqrt[6]{2}\left[\cos\left(\frac{\pi}{12}\right)+i\sin\left(\frac{\pi}{12}\right)\right]$$

$$\sqrt[6]{2}\left[\cos\left(\frac{3\pi}{4}\right)+i\sin\left(\frac{3\pi}{4}\right)\right]$$

$$\sqrt[6]{2}\left[\cos\left(\frac{5\pi}{6}\right)+i\sin\left(\frac{5\pi}{6}\right)\right]$$

Exercise

Find the four fourth roots of - 16 i.

Solution

First, we must write – 16 i in trigonometric form. The magnitude is 1, and the direction angle is π. So the trigonometric form of – 16 i is 16 (cos π + i sin π).

There will be four fourth roots. The magnitude of each of the fourth roots will be the fourth root of the magnitude of – 16 i. So the magnitude of each of the fourth roots will be 2. The angle of the first of the four fourth roots will be one-fourth of the angle for – 16 i, so the angle of the first of the four fourth roots will be π/4. The four fourth roots will lie evenly spaced around the unit circle, so they will lie an angular distance of π/2 radians apart. So the four angles associated with the four fourth roots will be π/4, 3 π/4, 5 π/4 and 7 π/4. So the roots will be:

2 [cos (π/4) + i sin (π/4)] = –2 + –2 i

2 [cos (3 π/4) + i sin (3 π/4)] = - –2 + –2 i

2 [cos (5 π/4) + i sin (5 π/4)] = - –2 - –2 i

2 [cos (7 π/4) + i sin (7 π/4)] = –2 - –2 i.

Trigonometry and Vectors

A vector in the plane is a directed line segment (represented by an arrow) from some initial point P to some terminal point Q. Two vectors are considered equivalent if they have the *same length* and the *same direction*. When the initial and terminal points are named with letters such as P and Q, the vector can be represented as PQ with an arrow over the top, i. e. $\overrightarrow{PQ}$. A *standard vector* is a vector in *standard position*, which means a vector with initial point at the origin in the Cartesian coordinate system. Every vector in the plane is equivalent to a standard vector.

Displacement is an example of a quantity measured by a vector. Suppose an object in a plane (the floor, for example) is moved three units to the east and four unit north. Then it would have been displaced a total of five units in an approximately north easterly direction along the diagonal of a right triangle with sides three and four. In the science of Physics, force is a quantity measured by a vector. Every force has a magnitude and a direction of application. The 'length' of a force vector equals the magnitude of the force. Consider how one would 'add' two displacements.

Suppose an object is moved 2 units west, then one unit south. Next, it is moved 3 units east and 4 units north. The sum or result of the two displacements would be the same as if the object had been moved 1 unit east and 3 units north. Picture the first displacement as an arrow going from the initial position of the object to its next position. Then think of the second displacement as an arrow going from the second position of the object to the third position. Then the resulting displacement would be an arrow going from the first position of the object to the third position of the object. See the diagram:

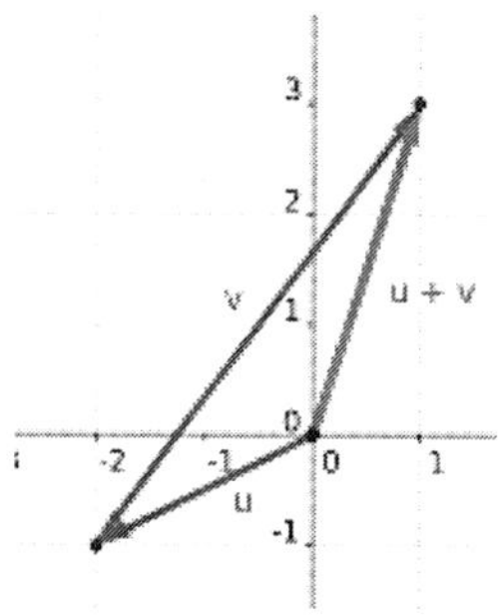

We see that the addition of vectors can be represented by placing the initial point of the second vector at the terminal point of the first vector, then the sum of the two vectors is the vector beginning at the initial point of the first vector and ending at the terminal point of the second vector. This is sometimes called the triangle rule for the addition of vectors.

Using this same example, we can introduce the idea of the *components* of a vector. Vector components are sometimes indicated by an ordered pair of numbers between two 'angle brackets'. Thus, the first displacement in this example—two units vest and one unit south—could be represented by the components < -2, -1 >. The second displacement—three units east and four units north—could be represented by the components < 3, 4 >. Notice that it is easy to find the components of the sum or resultant vector—just add the components of the vectors. < -2, -1 > + < 3, 4 > = < 1, 3 >.

Now, consider a standard vector v = < a, b >. (We'll denote vectors with lower case bold letters.) The length or magnitude of of v is denoted by | v | and computed using the Pythagorean Theorem.

$$|v| = \sqrt{a^2 + b^2}$$

Notice that *a*/| v | = sin θ and *b*/| v | = cos θ, where θ is the angle between v and the positive *x* axis.

So v = < *a*, *b* > = < | v | sin θ, | v | cos *t* > = | v | < sin θ, cos θ>. That is, v = | v | < sin θ, cos θ >.

The vector < sin θ, cos θ > is called the *direction* of v.

Notice that we factored out the | v | from the components of the vector. We can do this because multiplying a vector by a number changes the length of the vector accordingly.

If we multiply a vector by 2, the length is doubled and the direction is unchanged. If we multiply a vector by one-half, the length is halved and the direction is unchanged. However, if we multiply by - 1, the length is unchanged and the direction is reversed.

In general, if v = < a, b >, the *c* v = < *c a*, *c b* >

Exercise

Find the magnitude of the direction vector < sin θ, cos θ >.

Solution

The magnitude of the direction vector < sinθ, cosθ > equals the square root of the sum of the squares of the components. So the magnitude is 1. Direction vectors are sometimes called *unit* vectors because they have unit length.

Exercise

Given that v = < -2, 2 >, find the magnitude, the direction angle θ and the direction vector. Find the components of 3 v, and -2 v.

Solution

v = < -2, 2 >, | v | = 2"2

v is in quadrant II and tan θ = - 1, so θ = 135Ú. The direction vector is v/| v | = < -"2/2, –2/2 >.

3v = < -6, 6 > and – 2v = < 4, -4 >.

Two special unit vectors are sometimes defined:

i = < 1, 0 > corresponding to the direction angle 0° and

j = < 0, 1 > corresponding to the direction angle 90°.

Every vector in the plane can be represented as a linear combination of the unit vectors i and j.

$< a, b > = < a, 0 > + < 0, b > = a < 1, 0 > + b < 0, 1 > = a\,\mathrm{i} + b\,\mathrm{j}$

Thus, $a\,\mathrm{i} + b\,\mathrm{j}$ is an alternate way of writing a vector in component form.

Exercise

Find the components of a vector of magnitude 4 and direction angle 45°.

Solution

$|\mathrm{v}| = 4, \theta = 45^\circ$

$\mathrm{v} = |\mathrm{v}| < \cos 45^\circ, \sin 45^\circ > = \left\langle 2\sqrt{2}, 2\sqrt{2} \right\rangle$

Sometimes vectors are represented solely in terms of their magnitudes and directions, with specifying their components.

Consider the following problem:

Two vectors u and v of magnitude of 5 and 2 respectively are separated by an angle of 30°. Find the magnitude of their sum u + v and the angle the sum makes with the first vector.

The vectors u and v are initially drawn from the same initial point so that we can show the 30° separation.

A copy of u is then moved so that it begins at the terminal point of v. This new copy of u makes an angle of $180^\circ - 30^\circ = 120^\circ$ with v. This gives us a triangle with sides 2 and 5 and an included angle of 120°. Thus, we can find the magnitude of u + v using the Law of Cosines.

$|\mathrm{u} + \mathrm{v}|^2 = 5^2 + 2^2 - 2\,(5)\,(2) \cos 150^\circ = 29 + 20 \cos 30^\circ = 46.3205$ approximately.

So $|\mathrm{u} + \mathrm{v}|$ is approximately 6.806.

To find the angle θ the sum makes with u, note that it is congruent to the angle opposite v in the triangle. Thus using the Law of Cosines again,

$\cos \theta = 0.9891$. Thus, the angle is approximately 8.45°.

Exercise

Two vectors u and v of magnitude of 3 and 2 respectively are separated by an angle of 45°. Find the magnitude of their sum u + v and the angle the sum makes with the first vector.

Solution

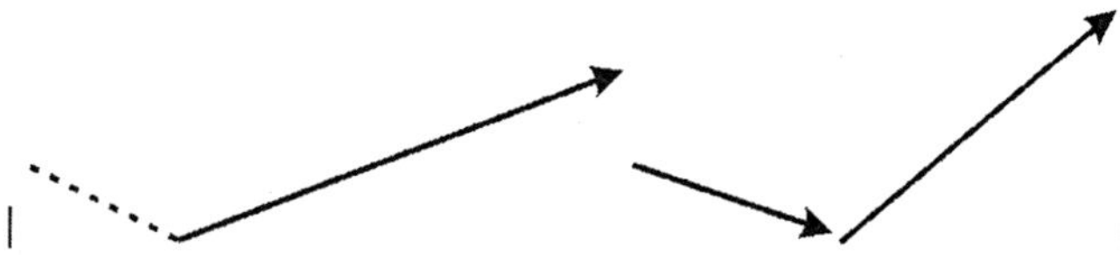

u | =3, | v | = 2

$\theta = 45°$

180° - 45° = 135° is the angle of the triangle opposite the side representing the sum of the vectors.

The angle opposite |v| is the angle between the vector u and the vector u + v.

$|u + v|^2 = |u|^2 + |v|^2 - 2\ |u|\ |v|\ \cos 135° =$

So, $|\ u + v\ | = 4.635$

Let ϕ denote the angle between u and u + v. Then using the Law of Cosines

$$\cos\varphi = \frac{|u|^2 + |u+v|^2 - |v|^2}{2|u||u+v|} \approx 0.9523$$

$$\varphi \approx 17.8°$$

Chapter 4

Uses of Matrix Algebra

Elementary Transformations

Elementary matrix transformations or Elementary row and column transformations are linear transformations which are normally used in gauss elimination to solve a set of linear equations.

We distinguish three types of elementary transformations and their corresponding matrices:

1. Row switching transformations,
2. Row multiplying transformations,
3. Linear combinator transformations.

Row Switching Transformations

This transformation, T_{ij}, switches all matrix elements on row i with their counterparts on row j. The matrix resulting in this transformation is:

$$Tij = \begin{bmatrix} 1 & & & & & & \\ & \ddots & & & & & \\ & & 0 & 1 & & & \\ & & & \ddots & & & \\ & & 1 & 0 & & & \\ & & & & & \ddots & \\ & & & & & & 1 \end{bmatrix},$$

Properties:

- The inverse of this matrix is itself: $T_{ij}^{-1}=T_{ij}$.
- When applied to a matrix A: $det[TA]=-det[A]$.
- The matrix and it's inverse are lower triangular matrices.

Row Multiplying Transformations

This transformation, $T_i(m)$, multiplies all elements on row i with m. The matrix resulting in this transformation is:

$$Ti(m)=\begin{bmatrix}1 & & & & & & \\ & \ddots & & & & & \\ & & 1 & & & & \\ & & & m & & & \\ & & & & 1 & & \\ & & & & & \ddots & \\ & & & & & & 1\end{bmatrix},$$

Properties:

- The inverse of this matrix is: $T_i(m)^{-1}=T_i(1/m)$.
- When applied to a matrix A: $det[TA]=m det[A]$.
- The matrix and it's inverse are lower triangular matrices.

Linear Combinatory Transformations

This transformation, $T_{ij}(m)$, substracts row i multiplied by m from row j. The matrix resulting in this transformation is:

$$Tij(m)=\begin{bmatrix}1 & & & & & & \\ & \ddots & & & & & \\ & & 1 & & & & \\ & & & \ddots & & & \\ & & & -m & 1 & & \\ & & & & & \ddots & \\ & & & & & & 1\end{bmatrix},$$

Properties

- The inverse of this matrix is: $T_{ij}(m)^{-1}=T_{ij}(-m)$.
- When applied to a matrix A: $det[TA]=det[A]$.
- The matrix and it's inverse are lower triangular matrices.

Rank of Matrix

Let,

$$A=\begin{bmatrix}a_{11} & a_{12} & \cdots & a_{1n} \\ a_{21} & a_{22} & \cdots & a_{2n} \\ \vdots & \vdots & \ddots & \vdots \\ a_{m1} & a_{m2} & \cdots & a_{mn}\end{bmatrix}.$$

The i'th row of A is,

$$row_i(A) = \begin{bmatrix} a_{i1} & a_{i2} & \cdots & a_{in} \end{bmatrix}, i = 1, 2..., m,$$

and the j'th column of A is,

$$col_j(A) = \begin{bmatrix} a_{1j} \\ a_{2j} \\ \vdots \\ a_{mj} \end{bmatrix}, j = 1, 2, ..., n.$$

Definition of Row Space and Column Space

$$span\{row_1(A), row_2(A), ..., row_m(A)\},,$$

which is a vector space under standard matrix addition and scalar multiplication, is referred to as *the row space.* Similarly,

$$span\{col_1(A), col_2(A), ..., col_n(A)\},,$$

which is also a vector space under standard matrix addition and scalar multiplication, is referred to as *the column space.*

Definition of Row Equivalence

A matrix B is row equivalent to a matrix A if B result from A via elementary row operations.

Example: Let,

$$A = \begin{bmatrix} 1 & 2 & 3 \\ 4 & 5 & 6 \\ 7 & 8 & 9 \end{bmatrix}, B_1 = \begin{bmatrix} 4 & 5 & 6 \\ 1 & 2 & 3 \\ 7 & 8 & 9 \end{bmatrix}, B_2 = \begin{bmatrix} 2 & 4 & 6 \\ 4 & 5 & 6 \\ 7 & 8 & 9 \end{bmatrix}, B_3 = \begin{bmatrix} 1 & 2 & 3 \\ 3 & 3 & 3 \\ 7 & 8 & 9 \end{bmatrix}$$

Since,

$$A = \begin{matrix} (1) \\ (2) \\ (3) \end{matrix} \begin{bmatrix} 1 & 2 & 3 \\ 4 & 5 & 6 \\ 7 & 8 & 9 \end{bmatrix} \xrightarrow{(1) \leftrightarrow (2)} B_1 = \begin{bmatrix} 4 & 5 & 6 \\ 1 & 2 & 3 \\ 7 & 8 & 9 \end{bmatrix},$$

$$A = \begin{matrix} (1) \\ (2) \\ (3) \end{matrix} \begin{bmatrix} 1 & 2 & 3 \\ 4 & 5 & 6 \\ 7 & 8 & 9 \end{bmatrix} \xrightarrow{(1) = 2*(1)} B_2 = \begin{bmatrix} 2 & 4 & 6 \\ 4 & 5 & 6 \\ 7 & 8 & 9 \end{bmatrix},$$

$$A = \begin{matrix}(1)\\(2)\\(3)\end{matrix}\begin{bmatrix}1 & 2 & 3\\4 & 5 & 6\\7 & 8 & 9\end{bmatrix} \xrightarrow{(2)=(2)-(1)} B_3 = \begin{bmatrix}1 & 2 & 3\\3 & 3 & 3\\7 & 8 & 9\end{bmatrix}, B_1, B_2, B_3$$

are all *row equivalent* to *A*.

Result: If A and B are two $m \times n$ row equivalent matrices, then the row spaces of A and B are equal.

Finding the Bases of the Row and Column Spaces

Suppose A is a $m \times n$ matrix. Then, the bases of the row and column spaces can be found via the following steps.

Step: Transform the matrix A to the matrix in reduced row echelon form.

Step:

- The nonzero rows of the matrix in reduced row echelon form form a basis of the row space of A.
- The columns corresponding to the ones containing the leading 1's form a basis. For example, if and the reduced row echelon matrix is

$$\begin{bmatrix}1 & \times & \times & \times & \times & \times\\0 & 0 & 1 & \times & \times & \times\\0 & 0 & 0 & 1 & \times & \times\\0 & 0 & 0 & 0 & 0 & 0\\\vdots & \vdots & \vdots & \vdots & \vdots & \vdots\\0 & 0 & 0 & 0 & 0 & 0\end{bmatrix},$$

then the 1'st, the 3'rd, and the 4'th columns contain a leading 1 and $col_1(A)$, $col_3(A)$, $col_4(A)$ thus form a basis of the column space of A.

Note: To find the basis of the column space is to find to basis for the vector space $span\{col_1(A), col_2(A), \ldots, col_n(A)\}$. Two methods introduced in the previous section can also be used. The method used in this section is equivalent to the second method in the previous section.

Example: Let,

$$A = \begin{bmatrix}1 & -2 & 0 & 3 & -4\\3 & 2 & 8 & 1 & 4\\2 & 3 & 7 & 2 & 3\\-1 & 2 & 0 & 4 & -3\end{bmatrix}.$$

Find the bases of the row and column spaces of A.

Solution:

Step: Transform the matrix A to the matrix in reduced row echelon form,

$$A=\begin{bmatrix} 1 & -2 & 0 & 3 & -4 \\ 3 & 2 & 8 & 1 & 4 \\ 2 & 3 & 7 & 2 & 3 \\ -1 & 2 & 0 & 4 & -3 \end{bmatrix} \xrightarrow{\text{in reduced row echelon form}} \begin{bmatrix} 1 & 0 & 2 & 0 & 1 \\ 0 & 1 & 1 & 0 & 1 \\ 0 & 0 & 0 & 1 & -1 \\ 0 & 0 & 0 & 0 & 0 \end{bmatrix}$$

Step:

- The basis for the row space is,

  $$\{[1 \quad 0 \quad 2 \quad 0 \quad 1],[0 \quad 1 \quad 1 \quad 0 \quad 1],[0 \quad 0 \quad 0 \quad 1 \quad -1]\}$$

- The columns corresponding to the ones containing the leading 1's are the 1'st, the 2'nd, and the 4'th columns. Thus,

$$\left\{\begin{bmatrix} 1 \\ 3 \\ 2 \\ -1 \end{bmatrix}, \begin{bmatrix} -2 \\ 2 \\ 3 \\ 2 \end{bmatrix}, \begin{bmatrix} 3 \\ 1 \\ 2 \\ 4 \end{bmatrix}\right\}$$

form a basis of the column space.

Definition of Row Rank and Column Rank

The dimension of the row space of A is called the row rank of A and the dimension of the column space of A is called the column rank of A.

Example (continue): Since the basis of the row space of A is $\{[1 \quad 0 \quad 2 \quad 0 \quad 1],[0 \quad 1 \quad 1 \quad 0 \quad 1],[0 \quad 0 \quad 0 \quad 1 \quad -1]\}$,
the dimension of the row space is 3 and the row rank of A is 3. Similarly,

$$\left\{\begin{bmatrix} 1 \\ 3 \\ 2 \\ -1 \end{bmatrix}, \begin{bmatrix} -2 \\ 2 \\ 3 \\ 2 \end{bmatrix}, \begin{bmatrix} 3 \\ 1 \\ 2 \\ 4 \end{bmatrix}\right\}$$

is the basis of the column space of A. Thus, the dimension of the column space is 3 and the column rank of A is 3.

Result: The row rank and column rank of the $m \times n$ matrix A are equal.

Definition of the rank of a matrix: Since the row rank and the column rank of a $m \times n$ matrix A are equal, we only refer to the rank of A and write rank (A).

Important result: If A is a $m \times n$ matrix, then,

$$\begin{aligned} &rank(A) + nullity(A) \\ &= \text{the dimension of the column space} + \text{the dimension of the null space} \\ &= n \end{aligned}$$

Example:

$$A = \begin{bmatrix} 1 & 0 & 0 & 0 & 0 \\ 0 & 1 & 0 & 0 & 0 \\ 0 & 0 & 1 & 0 & 0 \\ 0 & 0 & 0 & 0 & 0 \\ 0 & 0 & 0 & 0 & 0 \end{bmatrix}$$

and. n = 5.

Since,

$$\left\{ \begin{bmatrix} 1 \\ 0 \\ 0 \\ 0 \\ 0 \end{bmatrix}, \begin{bmatrix} 0 \\ 1 \\ 0 \\ 0 \\ 0 \end{bmatrix}, \begin{bmatrix} 0 \\ 0 \\ 1 \\ 0 \\ 0 \end{bmatrix} \right\}$$

is a basis of column space and thus rank (A) = 3. The solutions of Ax =0 are,

$$x_1 = 0,\ x_2 = 0,\ x_3 = 0,\ x_4 = s_1,\ x_5 = s_2,\ s_1, s_2 \in R.$$

Thus, the solution space (the null space) is,

$$s_1 \begin{bmatrix} 0 \\ 0 \\ 0 \\ 1 \\ 0 \end{bmatrix} + s_2 \begin{bmatrix} 0 \\ 0 \\ 0 \\ 0 \\ 1 \end{bmatrix} \Leftrightarrow span \left\{ \begin{bmatrix} 0 \\ 0 \\ 0 \\ 1 \\ 0 \end{bmatrix}, \begin{bmatrix} 0 \\ 0 \\ 0 \\ 0 \\ 1 \end{bmatrix} \right\}.$$

Then, $\begin{bmatrix} 0 \\ 0 \\ 0 \\ 1 \\ 0 \end{bmatrix}$ and $\begin{bmatrix} 0 \\ 0 \\ 0 \\ 0 \\ 1 \end{bmatrix}$ are the basis of the null space. and nullity (A) =2.

Therefore,

$$rank(A) + nullity(A) = 3 + 2 = 5 = n.$$

Important result: Let A be an matrix.

- A is nonsingular if and only if rank (A) = n.
- $rank(A) = n \Leftrightarrow A$ is nonsingular $\Leftrightarrow \det(A) \neq 0$
  $\Leftrightarrow Ax = b$ has a unique solution.
- $rank(A) < n \Leftrightarrow Ax = 0$ has a nontrivial solution.

Important result: Let A be an $m \times n$ matrix. Then,

$Ax = b$ has a solution rank (A) = rank [a | b]

Exercises

1. Determine the ranks of the coefficient and the augmented matrices as shown below.

$$1. \begin{bmatrix} -1 & 1 & 3 & 5 \\ 1 & 3 & 5 & 7 \\ 9 & 11 & 13 & 15 \\ -3 & -1 & 13 & \end{bmatrix} \qquad 2. \begin{bmatrix} 10 & 8 & 6 & 4 \\ 2 & 0 & -2 & -4 \\ -6 & -8 & -10 & -12 \\ -2 & -4 & -6 & -8 \end{bmatrix}$$

2. Let A be an $n \times n$ matrix with rank $(A) = n$ .Then prove that A is row-equivalent to I_n.
3. If P and Q are invertible matrices and PAQ is defined then show that Rank (PAQ) = rank (A).
4. Find matrices P and Q which are product of elementary matrices such that $B = PAQ$ where,

$$A = \begin{bmatrix} 2 & 4 & 8 \\ 1 & 3 & 2 \end{bmatrix} \text{ and } A = \begin{bmatrix} 1 & 0 & 0 \\ 0 & 1 & 0 \end{bmatrix}$$

5. Let A and B be two matrices. Show that
   1. If $A + B$ is defined, then rankl $(A + B) \leq$ rank (A) + rank (B),
   2. If AB is defined, then and rank $(AB) \leq$ rank (B)
6. Let A be any matrix of rank τ .Then show that there exists invertible matrices B_i, C_i such that

$$B_1 A = \begin{bmatrix} R_1 & R_2 \\ 0 & 0 \end{bmatrix}, AC_1 = \begin{bmatrix} S_1 & 0 \\ S_3 & 0 \end{bmatrix}, B_2 A C_2 = \begin{bmatrix} A_1 & 0 \\ 0 & 0 \end{bmatrix},$$

and $B_3AC_3 = \begin{bmatrix} I_r & 0 \\ 0 & 0 \end{bmatrix}$. Also, prove that the matrix A_1 is an $\tau \times \tau$ invertible matrix.

7. Let A be an $m \times n$ matrix of rank t .Then A can be written as $A = BC$ where both B and C have rank t and B is a matrix of size $m \times \tau$ and C is a matrix of size $\tau \times n$.
8. Let A and B be two matrices such that AB is defined and rank (A) = rank (AB) .Then show that $A = ABX$ for some matrix x Similarly, if BA is defined and rank (A) = rank (BA), then $A = YBA$ for some matrix Y.[Hint: Choose non-singular matrices P, Q and R such that,

$$PAQ = \begin{bmatrix} A_1 & 0 \\ 0 & 0 \end{bmatrix} \text{ and}$$

$$P(AB)R = \begin{bmatrix} C & 0 \\ 0 & 0 \end{bmatrix}. \text{ Define } X = R\begin{bmatrix} C^{-1}A_1 & 0 \\ 0 & 0 \end{bmatrix}Q^{-1} \text{]}$$

9. If matrices B and C are invertible and the involved partitioned products are defined, then show that,

$$\begin{bmatrix} A & B \\ C & 0 \end{bmatrix}^{-1} = \begin{bmatrix} 0 & C^{-1} \\ B^{-1} & -B^{-1}AC^{-1} \end{bmatrix}.$$

10. *Suppose A is the inverse of a matrix B Partition A and B as follows:*

$$A = \begin{bmatrix} A_{11} & A_{12} \\ A_{21} & A_{22} \end{bmatrix}, B = \begin{bmatrix} B_{11} & B_{12} \\ B_{21} & B_{22} \end{bmatrix}.$$

If A_{11} is invertible and $P = A_{22} - A_{21}\left(A_{11}^{-1}A_{12}\right)$, then show that:

$$B = A_{11}^{-1} + \left(A_{11}^{-1}A_{12}\right)P^{-1}\left(A_{21}A_{11}^{-1}\right), B = -P^{-1}\left(A_{21}A_{11}^{-1}\right), B_{12}\left(A_{11}^{-1}A_{12}\right)P^{-1},$$

and $B_{22} = P^{-1}$

Inverse of Matrix by Elementary Transformation

Linear algebra is perhaps the most powerful and useful math course you can ever take in your entire academic career. Resting at the heart of this course is something called a matrix. A matrix is a rectangular array of numbers organized in rows and columns and encased in brackets. Matrices are very useful due to the fact that they can be easily manipulated. In this page we will explore how to find the inverse of a matrix and its uses. Much like normal numbers we use the notation A^{-1} to denote the inverse of matrix A. Some important things to remember about inverse matrices is they are not commutative, and a full generalization is possible only if the matrices you are using a square. (meaning they have the same number of rows and columns, an n x n matrix) An n x n matrix (A) is said to be invertible if there is an n x n matrix (C) such that CA= I and AC= I where I is the n x n identity matrix. An identity matrix is a square matrix with ones on the diagonal and zeros elsewhere. 3 × 3 identity matrix

$$\begin{bmatrix} 1 & 0 & 0 \\ 0 & 1 & 0 \\ 0 & 0 & 1 \end{bmatrix}$$

Basically A^{-1} A= I and A A^{-1} = I where A is an invertible matrix and A^{-1} is the inverse of A. A matrix is said to be a singular matrix if it is non-invertible. A matrix is a nonsingular matrix if it is an invertible matrix.

A simple formula for finding the inverse of a 2 × 2 matrix is given by Theorem 4:

$$A^{-1} = \begin{bmatrix} a & b \\ c & d \end{bmatrix} = \frac{1}{ad - bc} \begin{bmatrix} d & -b \\ -c & a \end{bmatrix}$$

We call the quantity ad–bc the determinant of the matrix. (det A = ad–bc) A 2 × 2 matrix is invertible if and only if (iff) its determinant does not equal 0.

$$\begin{bmatrix} 3 & 4 \\ 5 & 6 \end{bmatrix}^{-1} = \frac{1}{-2} \begin{bmatrix} 6 & -4 \\ -5 & 3 \end{bmatrix} = \begin{bmatrix} -3 & 2 \\ 5/2 & -3/2 \end{bmatrix}$$

Theorem 5 reveals something else useful about the inverse of matrices. Theorem 5 states that if matrix A is invertible then the equation Ax = b has a unique solution, x. We can find this solution by x = A^{-1} b. The following example demonstrates this usefulness of this equation. The following proof will help prove theorem 5 by proving 1) that the solution exists and 2) this solution is unique.

1. let x = A^{-1} b, plug this in to check

   A(x) = A A^{-1} b= Ib = b
2. We will prove this portion of theorem 5 by contradiction. Remember A(x + y) = Ax = Ay

AB does not equal BA, so assume there are two different solutions x does not equal y with Ax = b and Ay = b, but then Ax–Ay = 0, A(x-y) = 0 Multiply by A^{-1}, A^{-1} 0 = A^{-1} A(x-y) = I(x-y) = x-y x-y = 0, therefore x = y. However we said in the beginning that x does not equal y, this is a contradiction.

Example of solving Ax=b by using the inverse of A:

$$\begin{cases} 3x_1+4x_2=3 \\ 5x_1+6x_2=7 \end{cases} \Rightarrow Ax=b \Rightarrow x=A^{-1}b=\begin{bmatrix} -3 & 2 \\ 5/2 & -3/2 \end{bmatrix}\begin{bmatrix} 3 \\ 7 \end{bmatrix}=\begin{bmatrix} 5 \\ -3 \end{bmatrix}$$

More useful properties of inverse matrices are revealed in Theorem 6. This theorem states: If A is an invertible matrix, then $A^{-1}$is invertible and (A^{-1}) –1 = A,

If A and B are n x n invertible matrices, then so is AB and the inverse of AB is the product of the inverses of A and B in the reverse order. (AB)–1 = B–1 A^{-1}

If A is an invertible matrix, then so is A^T , and the inverse of A^T is the transpose of A^{-1} . That is (A^T)–1 = $(A^{-1})^T$ Are you skeptical at all with theorem 6? Well, incase you are, here are the proofs for each part.

To Prove 1: we need to find matrix C so that A^{-1} C = I and C A^{-1} = I We already know that these equations will still hold true if we put A in place of C. (see above) Thus A^{-1} is invertible, and A is its inverse.

To Prove 2: we first must calculate (AB)(B–1 A^{-1}) = A(BB–1) A^{-1} = AI A^{-1} = A A^{-1} = I a similar calculation shows that (B–1 A^{-1})(AB) = I.

$(A^{-1})^T A^T = (A A^{-1})^T - I^T - I$. And thus: $A^T (A^{-1})^T = I^T = I$. Thus we have proved that A^T is invertible, and its inverse is $(A^{-1})^T$

Also, it is useful to remember that the product of n x n invertible matrices is invertible, and the inverse is the product of their inverses in the reverse order.

Elementary Matrices

The usefulness of matrices continues to expand with the introduction of elementary matrices. An elementary matrix is a matrix that is obtained by performing a single elementary row operation to an identity matrix.

An elementary row operation is the process of either:

- Replacing one row of a matrix with the sum of itself and a multiple of another row
- Interchanging two rows
- Multiplying all entries in a row by a nonzero constant.

If an elementary row operation is performed on an n x n matrix A, the resulting matrix can be written as E A, where the n x n matrix E is created by performing the same row operation on I_m . The following example demonstrates this concept:

Row replacement:

$$\begin{bmatrix} 1 & 0 & 0 \\ 0 & 1 & 0 \\ -4 & 0 & 1 \end{bmatrix}\begin{bmatrix} a & b & c \\ d & e & f \\ g & h & i \end{bmatrix} = \begin{bmatrix} a & b & c \\ d & e & f \\ g-4a & h-4b & i-4c \end{bmatrix}$$

Interchange of rows:

$$\begin{bmatrix} 0 & 1 & 0 \\ 1 & 0 & 0 \\ 0 & 0 & 1 \end{bmatrix}\begin{bmatrix} a & b & c \\ d & e & f \\ g & h & i \end{bmatrix} = \begin{bmatrix} d & e & f \\ a & b & c \\ g & h & i \end{bmatrix}$$

Multiplication by a constant:

$$\begin{bmatrix} 1 & 0 & 0 \\ 0 & 1 & 0 \\ 0 & 0 & 5 \end{bmatrix}\begin{bmatrix} a & b & c \\ d & e & f \\ g & h & i \end{bmatrix} = \begin{bmatrix} d & e & f \\ a & b & c \\ 5g & 5h & 5i \end{bmatrix}$$

It should also be noted that each elementary matrix E is invertible. The inverse of E is the elementary matrix of the same type that transforms E back into I.

$$E^{-1}\begin{bmatrix} 1 & 0 & 0 \\ 0 & 1 & 0 \\ -4 & 0 & 1 \end{bmatrix}^{-1}\begin{bmatrix} 1 & 0 & 0 \\ 0 & 1 & 0 \\ +4 & 0 & 1 \end{bmatrix}$$

Finally theorem 7 gives us a way to visualize an inverse matrix and helps us develop a method of finding inverse matrices. Theorem 7 says that an n x n matrix, called A is invertible iff (if and only if) A is row equivalent to I_n , and any sequence of elementary row operations that reduces A to I_n also transforms I_n into A^{-1}.

Normal Form of a Matrix

The Singular Value Decomposition (SVD) of a matrix A is possibly the most powerful way of revealing the properties of this matrix.

The *normal form* (or "identity decomposition") of a matrix is another useful decomposition with a somewhat less ambitious goal : the only feature of A that we want to reveal is its *rank* (whereas SVD uncovers much more than that).

- Revealing the rank of a matrix
- Let A be any matrix mxn of rank r.

We will show that there exist :

- Two square nonsingular matrices $B(m\text{x}m)$ and $C(n\text{x}n)$,
- And a "pseudo-identity" matrix $H(m\text{x}n)$ whose off diagonal elements are 0s, but which is not square.
- The upper part of the diagonal (red strip in the illustration below) in H contains only 1s, and we'll show that their number is r, the rank of A.
- The lower part of the diagonal (gray strip) contains $(n–r)$ 0s, just as the rest of H.

Such that:

$A = B^{-1}HC^{-1}$ or, equivalently $H = BAC$

This decomposition may be represented as:

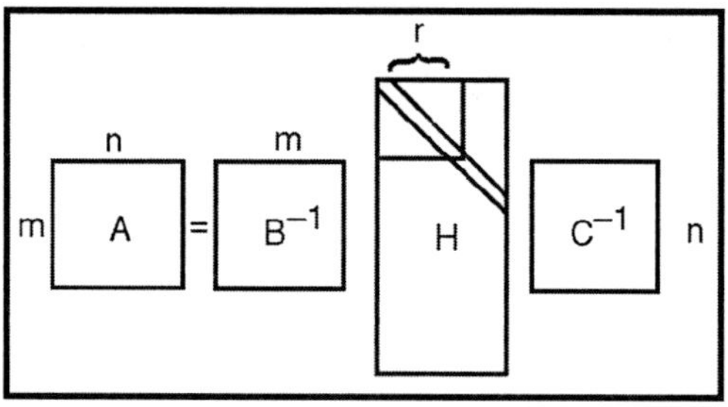

This figure illustrates the case $r < n < m$, but the decomposition is also valid for m n, and for $r = \min(m, n)$ (that is when A is full rank).

Despite the formal similarity with SVD, the normal form reveals less information about the matrix than SVD, as:

- Only the rank of A is read off the normal form,
- The matrices B and C are definitely not unique, and contain no straightforwardly usable information about A.

Elementary transformations:

On the other hand, the normal form is easier to elaborate than SVD. We'll show that B and C can be obtained from A by resorting only to so-called*elementary transformations* that come in three kinds :

- (R_1) Interchange of two rows.
- (R_2) Multiplication of a row by a (non zero) number.
- (R_3) Addition of the result of (R_2) to another row. with a similar set of elementary transformations for columns. These transformations can be represented by *elementary transformation matrices* that:
- Left-multiply A for row transformations, and
- Right-multiply A for column transformations.

Reduced normal form:

Just as SVD has a "full" and a "reduced" version, we'll show that the "full" normal form described above is equivalent to the following "reduced" form:

- Any matrix $A(mxn)$ of rank r can be factored into two full rank matrices F and G of rank r:

  $A = FG$

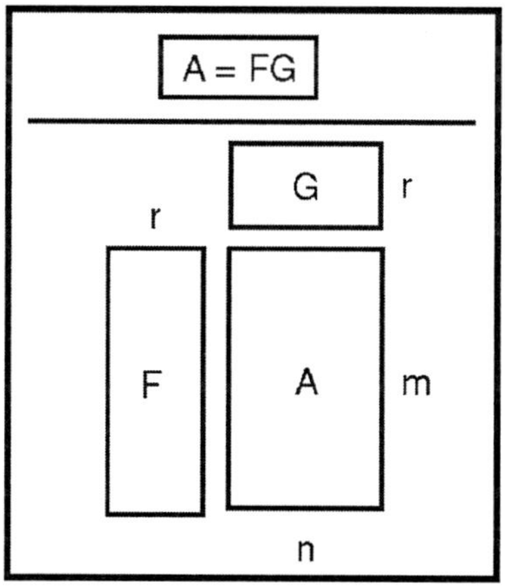

- F has m rows and G has n columns, so their product is mxn.
- Both are r columns (resp. rows) thick. So now the rank of A reads as the "thickness" of F or of G.

Consistency of System of Linear Equations

In mathematics, a system of linear equations is a collection of two or more linear equations with the same set of variables in all the equations. In other words, we can say a system of linear equations is nothing but two or more equations that are being solved simultaneously.

Mostly, the system of equations can be used by the business people to predict their future events. They will model a real world situation in to system of equations to find the solution and manage their business. We can make an accurate prediction by using system of equations.

The solution of the system of equations is an ordered pair that satisfies each equation in the system.

Consider the two equations:

1. $x + y = 2$
2. $x - y = 2$

It forms a system of equations in two variables. The solution of this system is the ordered pair (x, y).

Consider the equations given below.

- $-3a + 2b - 6c = 6$
- $5a + 7b - 5c = 6$
- $a + 4b - 2c = 8$

It forms a system of equations in three variables. The solution of this system is the ordered pair (a, b, c).

The solution of a system of linear equations can be of three types.

They are:

1. One solution
2. Infinite solution and
3. No solution

If the graph of the equations intersects each other at a point, then ordered pair corresponding to the point of intersection will be the solution to that system. In this case, the system will have exactly *one solution.* If a system has exactly one solution, then the equations are said to be *independent.*

If the graph of the equations coincides, then all the points on the line will be the solution to that system. In this case, the system will have *infinite number of solutions.* If the system has infinite number of solutions, then the equations are said to be *dependent.* If the graphs of the equations are parallel, then the system of equations will have *no solution.* Because parallel lines never intersect each other. If the system has at least one solution (one solution or infinitely many solutions), then it is said to be *consistent system.* If the system has no solution, then it is said to be *inconsistent system.* The following figure will give clear picture of what we have learnt above.

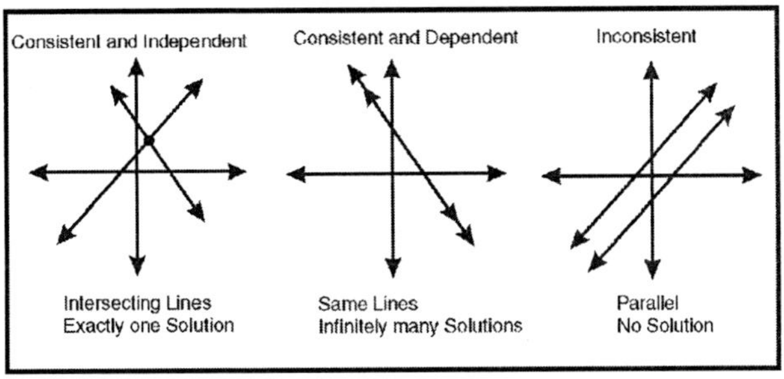

Systems of Equations

When you are trying to calculate the solution of a system of linear equations, you can will arrive at one of three distinct cases: These cases only apply to systems of two lines. If you are working with systems with three or more linear equations (lines), you cannot use the blanket generalizations made below.

The system has exactly 1 solution:

- Systems have 1 and only 1 solution when the two lines have different slope. Think about it, if the two lines have different slopes then eventually at some point they must meet. After all the lines are not parallel.

System has no solutions:

- Systems have no solution when the lines are parallel (ie have the same slope) and the lines have different y-intercepts.

As an example look at the following two lines:

1. Line 1: $y = 5x + 13$
2. Line 2: $y = 5x + 12$

The system has infinite solutions:

- Systems have infinite solutions when the lines are parallel and the lines have the same y-intercept. If two lines have the same slope (ie are parallel) and the same y-intercept, they are actually the same exact line. In other words, systems have infinite solutions when the two lines are the same line!

As an example consider the following two lines:

1. Line 1: $y = x + 3$
2. Line 2: $2y = 2x + 6$

These two lines are exactly the same line. If you multiply line 1 by two you get line 2.

Methods of Finding Solution of Linear Equations

Graph: By looking at where lines intersect (meet) on a graph On the left, the system of linear equations is the following two lines:

1. y = x + 1
2. y=2x

What is the solution?

Answer: The point (1,2) is where the two lines intersect.

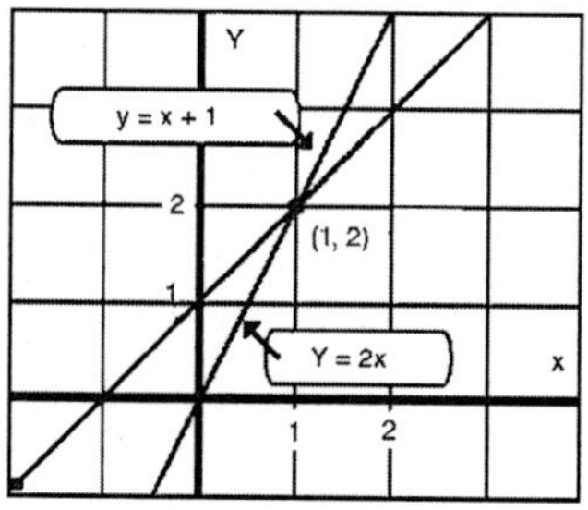

Example problems: Use the graph method to solve the system of equations on the left.

- y = 2x+1
- y = 4x–1

Answer: The solution of this system is the point of intersection: (1,3).

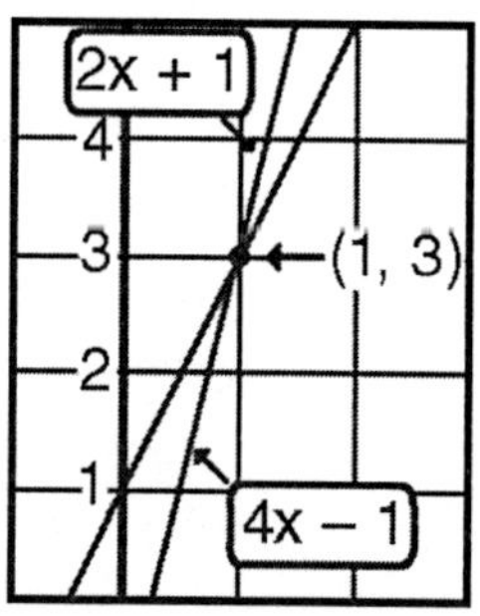

Solve the following system of linear equations by graphing.

- 2y = 4x + 2

and,

- 2y = –x + 7

Answer:

- $2y = 4x + 2$
  - $\frac{1}{2}(2y) = \frac{1}{2}(4x + 2)$
  - $y = 2x + 1$
- $2y = 8x-2$
  - $\frac{1}{2}(2y) = \frac{1}{2}(8x-2)$
  - $y = 4x - 1$

This system of lines is the same system that we looked at in the last example. Therefore, the solution is the point (1,3).

Algebraic equation: by setting the equations of the system equal to each other then solving this equation.

Let's take another look at the system of equations from above:

- $y=2x+1$
- $y=4x-1$

By examining the graph we can see that the point of intersection, or the solution, is the point (1,3) where the lines intersected.

Steps for the algebraic method:

- Make sure that each linear equation is reduced to slope intercept form
  - (ie $y=3x+2$ is good but $2y=6x+4$ is NOT)
- Set the two equations equal to each other
  - $2x+1=4x-1$
- Solve for X
  - $2x+1=4x-1$
  - $2=2x$
  - $x= 1$
- Insert x value into either equation to determine y coordinate of solution
  - $4(1)-1=3$
- The solution is the ordered pair you've just calculated
  - (1,3)

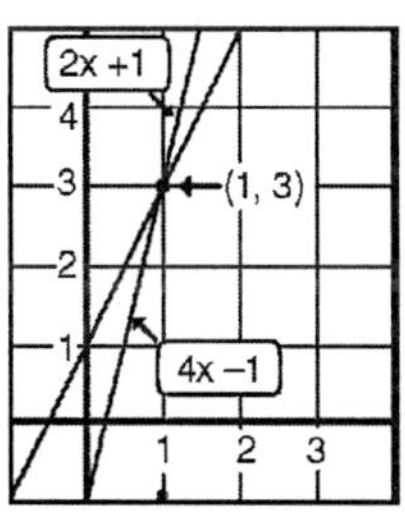

Practice Problem

What is the solution to the following system of linear equations:

- Line 1: y=3x – 1
- Line 2: y= x – 5

Step: Set the Two Equations Equal to each other then solve for x,

$$\begin{array}{l} 3x-1=x-5 \\ \quad +1 \qquad +1 \\ \hline 3x=x-4 \\ \quad -x \quad -x \\ \hline 2x \quad =-4 \\ 2x/2=-4/2 \\ x=-2 \end{array}$$

Step

1. Substitute the x value, –2, into either equation to determine y coordinate of solution
2. y=(– 2)–5 →–7
3. The solution is the point (– 2, –7)

Practice Problem

Set the two equations equal to calculate the solution to the system below:

- Line 1: y = x + 1
- Line 2: y = 2x

Answer:

$$\begin{array}{c} y=x+1 \qquad y=2x \\ x+1=2x \\ -x \qquad -x \\ \hline 1=x \\ \downarrow \end{array}$$

$$\begin{array}{l} y=2x \\ y=2(1)=2 \\ Solution:(1,2) \end{array}$$

The solution of this system is the point of intersection: (1,2). Below is the graph showing this solution.

Practice problem:

Solve the system of linear equations by setting their equations equal:

- Line 1: y= x +5
- Line 2: y= 2x +2

Answer:

$$y = x + 5 \qquad y = 2x + 2$$

$$x + 5 = 2x + 2$$

$$\underline{\quad -x \qquad -x \quad}$$

$$5 = x + 2$$

$$\underline{\quad -2 \qquad -2 \quad}$$

$$3 = x$$

$$\downarrow$$

$$y = 2x + 2$$

$$y = 2(3) + 2 = 8$$

$$Solution:(3,8)$$

The solution of this system is the point of intersection: (3,8). Below is the graph showing this solution.

Practice problem:

Solve the system of linear equations by setting their equations equal:

- Line 1: y= x – 1
- Line 2: y= 2x +2

Answer:

$$y = x - 1 \qquad y = 2x + 2$$

$$x - 5 = 2x + 2$$

$$\underline{\quad -x \qquad -x \quad}$$

$$-1 = x + 2$$

$$\underline{\quad -2 \qquad -2 \quad}$$

$$3 = x$$

$$\downarrow$$

$$y = 2x + 2$$

$$y = 2(-3) + 2 = -4$$

$$Solution:(-3,-4)$$

The solution of this system is the point of intersection: (– 3,– 4).

Practice problem:

Solve the system of linear equations by setting their equations equal:

- Line 1: y= x + 2
- Line 2: y= x + 8

Answer:

$$y = x + 2 \quad y = x + 8$$

$$x + 2 = x = 8$$

$$\frac{-x \quad -x}{2 \neq 8}$$

This system of linear equation has no solution Substitution: by solving for one of the variables and substituting its value in to the other equation. The substitution method involves algebraic substitution of one equation into a variable of the other.

Refresher: Substitution.

- Equation 1: x = 5
- Equation 2: y = x +2

How to substitute:

- Use equation 1(x = 5) to substitute 5 for x in second equation
  - y = (5) + 2
- So love for Y
  - y = 5 + 2 = 7

Substitution example two:

- *Line* 1 : y=2x+1
- *Line* 2 : 2y=3x–2
- *Step* 1: Substitute one equation into the other
  - 2(2x + 1)=3x–2
- *Step* 2: Now that you have a single variable equation, solve for that variable's equation
  - 4x + 2 = 3x–2
  - x + 2=–2
  - x= – 4
- *Step* 3 : Once you have solved for the one variable insert that variable back into either equation to obtain the value of y at the solution.

– Insert x= –4 to find y value → y = 2(–4)+1= –7

This example's solution is (–4, –7).

Let's examine the graph of this system to see if we correctly solved the problem.

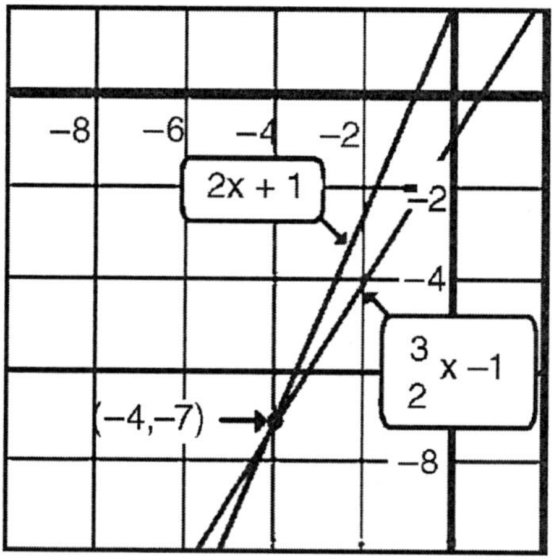

Practice problem:

Use the substitution method to solve the system:

- Line 1: y= x + 1
- Line 2: y= 2x

Answer:

$$y = \underbrace{x+1} \quad 2y = 3x$$

$$\downarrow$$

$$2y = 3x$$

$$2(x+1) = 3x$$

$$2x + 2 \ = 3x$$

$$\underline{-2x \qquad -2x}$$

$$2 = x$$

$$\downarrow$$

$$y = 3x = 2(3) = 6$$

$$Solution: (2,6)$$

The solution of this system is (2,6).

Practice problem:

Use the substitution method to solve the system:

- Line 1: y = 5x – 1
- Line 2: y= 3x + 12

Answer:

$$y = \underbrace{5x-1}_{\downarrow} \quad 2y = 3x+12$$

$$2y = 3x+12$$

$$2(5x-1) = 3x+12$$

$$10x-2 \quad = 3x+12$$

$$\underline{-3x \qquad\qquad -3x}$$

$$7x-2 = 12$$

$$\underline{\quad +2 \quad +2}$$

$$7x = \quad 14$$

$$\frac{7x}{7} = \frac{14}{7}$$

$$x = 2$$

$$\downarrow$$

$$y = 5x-1$$

$$y = 5(2)-1$$

$$y = 9$$

$$Solution: (2,9)$$

This system of lines has a solution at the point (2,9).

Practice Problem Four:

Use substitution to solve the system:

- Line 1: y = 3x + 1
- Line 2: 4y = 12x + 4

Answer:

$$y = \underbrace{3x+1}_{\downarrow} \quad 4y = 12x+4$$

$$4y - 12x + 4$$

$$4(3x+1) = 12x+4$$

$$12x+4 \quad = 12x+4$$

This system has an infinite number of solutions.

Practice problem:

Use substitution to solve the system:

- Line 1: y = 3x + 1
- Line 2: 4y = 12x + 3

Answer:

$$y = \underbrace{3x+1}_{\downarrow} \quad 4y = 12x + 3$$

$$4y = 12x + 3$$
$$4(3x+1) = 12x + 3$$
$$12x + 4 = 12x + 3$$
$$\underline{-12x \quad = -12x}$$
$$4 = 3$$
$$No\,solution$$

Whenever you arrive at a contradiction such as 3 = 4, your system of linear equations has no solutions. When you use these methods (substitution, graphing etc...) to find the solution what you're really asking is at what point (which x coordinate and y coordinate) do the two lines intersect. However, if the only time that they intersect is when 3 = 4, they are never going to intersect since 3 does not equal four! This system has an no solutions.

Elimination: Elimination involves algebraic mani- pulations of two or more equations. The end goal is to eliminate a variable by creating opposite coefficients Elimination method is an algebraic method for solving systems. To use elimination you perform an operation on 1 equation then add the two equations so that one of the variables cancels.

Example of elimination:

- Line 1: y = x + 1 • Line 2: y = – x

$$y = x + 1 \qquad y = -x$$

$$\left.\begin{array}{l} y = x + 1 \\ \underline{+y = -x} \\ 2y = 1 \\ y = \dfrac{1}{2} \end{array}\right\} E\lim inat\, x$$

$$\downarrow$$

$$y = -x\}\, substitute$$

$$-\frac{1}{2} = x$$

$$solution\left(-\frac{1}{2}, \frac{1}{2}\right)$$

Linear and Orthogonal Transformations

Given a form on a vector space V represented by the transformation B and a linear map T: $V \to V$, we've seen how to transform B by the action of T. That is, the space of all bilinear forms is a vector space which carries a representation of $GL(V)$. But given a particular form B, what is the stabilizer of B? That is, what transformations in $GL(V)$ send B back to itself. Before we answer this, let's look at it in a slightly different way. Given a form B we have a way of pairing vectors V in to get scalars. On the other hand, if we have a transformation T we could use it on the vectors before pairing them. We're looking for those transformations so that for every pair of vectors the result of the pairing by B is the same before and after applying T. So let's look at the action we described last time: the form B is sent to T^*BT. So we're looking for all T so that $T^* BT = B$

We say that such a transformation is B-orthogonal, and the subgroup of all such transformations is the "orthogonal group" $O(V,B) \subseteq GL(V)$. Sometimes, since the vector space V is sort of implicit in the form B, we abbreviate the group to $O(B)$. Now there's one particular orthogonal group that's particularly useful. If we've got an inner-product space V (the setup for having our bra-ket notation) then the inner product itself is a form, and it's described by the identity transformation. That is, the orthogonality condition in this case is that $T^*T = I_v$

A transformation is orthogonal if its adjoint is the same as its inverse. This is the version of orthogonality that we're most familiar with. Commonly, when we say that a transformation is "orthogonal" with no qualification about what form we're using, we just mean that this condition holds.

Let's take a look at this last condition geometrically. We use the inner product to define a notion of (squared-)length $\langle v|v\rangle$ and a notion of (the cosine of) angle $\langle w|v\rangle$. So let's transform the space by T and see what happens to our inner product, and thus to lengths and angles.

$$\langle T(w)|T(v)\rangle = \langle w|T^*Tv\rangle$$

First off, note that no matter what T we use, the transformation in the middle is self-adjoin and positive-definite, and so the new form is symmetric and positive-definite, and thus defines another inner product. But when is it the *same* inner product? When $T^*T = I_v$, of course! For then we have,

$$\langle T(w)|T(v)\rangle = \langle w|T*T|v\rangle = \langle w|Iv|v\rangle = \langle w|v\rangle$$

So orthogonal transformations are exactly those which preserve the notions of length and angle defined by the inner product. Geometrically, they correspond to rotations and reflections that change orientations, but leave lengths of vectors the same, and leave the angle between any pair of vectors the same.

Eigen Values and Eigen Vectors

Consider the square matrix A. We say λ that is an eigenvalue of A if there exists a non-zero vector x such that $Ax = \lambda x$. In this case, x is called an eigenvector (corresponding to λ), and the pair (λ,x) is called an eigenpair for A.

Let's look at an example of an eigenvalue and eigenvector. If you were asked if,

$$x = \begin{bmatrix} 1 \\ -2 \end{bmatrix}$$

is an eigenvector corresponding to the eigenvalue$\lambda = 0$ for

$$x = \begin{bmatrix} 6 & 3 \\ -2 & -1 \end{bmatrix},$$

you could find out by substituting x, and A into the equation $Ax = \lambda x$.

$$Ax = \lambda x$$

$$\begin{bmatrix} 6 & 3 \\ -2 & -1 \end{bmatrix}\begin{bmatrix} 1 \\ -2 \end{bmatrix} = 0\begin{bmatrix} 1 \\ -2 \end{bmatrix}$$

$$\begin{bmatrix} 0 \\ 0 \end{bmatrix} = \begin{bmatrix} 0 \\ 0 \end{bmatrix}$$

Therefore, λ and x are an eigenvalue and an eigenvector, respectively, for A.

Now that you have seen an eigenvalue and an eigenvector, let's talk a little more about them. Why did we require that an eigenvector not be zero? If the eigenvector was zero, the equation $Ax = \lambda x$ would yield $0 = 0$. Since this equation is always true, it is not an interesting case. Therefore, we define an eigenvector to be a non-zero vector that satisfies $Ax = \lambda x$.However, as we showed in the previous example, an eigenvalue can be zero without causing a problem. We usually say that x is an eigenvector corresponding to the eigenvalue λ if they satisfy $Ax = \lambda x$.

Since each eigenvector is associated with an eigenvalue, we often refer to an x and λ that correspond to one another as an eigenpair. Did you notice that we called x "an" eigenvector rather than "the" eigenvector corresponding to λ? This is because any non-zero, scalar multiple of an eigenvector is also an eigenvector. If you let c represent a scalar, then we can prove this fact through the following steps.

$A(cx) = cAx = c\lambda x = \lambda(cx)$

Since any non-zero, scalar multiple of an eigenvector is also an eigenvector, $\begin{bmatrix} 2 \\ -4 \end{bmatrix}$ and $\begin{bmatrix} -1 \\ 2 \end{bmatrix}$ are also eigenvectors corresponding $1 = 0$ to when,

$$A = \begin{bmatrix} 6 & 3 \\ -2 & -1 \end{bmatrix}.$$

You have already computed eigenvectors in this course. When we studied Markov chains, you computed an eigenvector corresponding to $A^T$when you found the matrix to which the probabilities seemed to converge after many steps. Any row of that matrix is an eigenvector for A^Tbecause all the rows of that matrix are the same. We write that row as a column vector when we use it as an eigenvector. The eigenvector that you found is called the dominant eigenvector.

Definition The dominant eigenvector of a matrix is an eigenvector corresponding to the eigenvalue of largest magnitude (for real numbers, largest absolute value) of that matrix. Although we only found one eigenvector, we found a very important eigenvector. Many of the "real world" applications are primarily interested in the dominant eigenpair. The method that you used to find this eigenvector is called the power method. The power method will be explained later in this chapter. An eigenvector corresponding to the transpose of a transition matrix is the transpose of any row of the matrix that A^kconverges to as k grows, these rows are all the same. The dominant eigenvalue is always 1 for a transition matrix. Let's look at the example that we used in the Markov chain chapter. Consider the matrix

$$A = \begin{bmatrix} .3 & .3 & .4 \\ .4 & .4 & .2 \\ .5 & .3 & .2 \end{bmatrix}.$$

If k is large,

$$A^k \approx \begin{bmatrix} .3\overline{8} & .3\overline{3} & .2\overline{7} \\ .3\overline{8} & .3\overline{3} & .2\overline{7} \\ .3\overline{8} & .3\overline{3} & .2\overline{7} \end{bmatrix}.$$

Therefore, an eigenvector corresponding to the dominant eigenvalue, $\lambda = 1$, is

$$\begin{bmatrix} .3\overline{8} \\ .3\overline{3} \\ .2\overline{7} \end{bmatrix}.$$

Let's see if $A^T x = \lambda x$ holds true.

$$\begin{bmatrix} .3 & .4 & .5 \\ .3 & .4 & .3 \\ .4 & .2 & .2 \end{bmatrix} \begin{bmatrix} .3\overline{8} \\ .3\overline{3} \\ .2\overline{7} \end{bmatrix} = 1 \begin{bmatrix} .3\overline{8} \\ .3\overline{3} \\ .2\overline{7} \end{bmatrix}.$$

Yes, the equation holds, so we have found an eigenpair corresponding to the transpose of the transition matrix.

Remark: For a transition matrix, the dominant eigenvalue is always 1. An eigenvector corresponding to $\lambda = 1$ for A^T is the transpose of any row of A^* where A^* is the matrix to which A^k converges as k grows. The matrix for which we are finding an eigenpair must have been set up so that the columns (not the rows) add to 1 for the eigenvector to be read from A^*; this is why we are dealing with A^T instead of A. These rules will require some modification if we are not dealing with a transition matrix. If we do not have a transition matrix, can we still use the power method? Yes we can, but we need to modify the steps a bit because the dominant eigenvalue will not necessarily be the number one. Let us explain how to use the power method. An example follows the remarks to help clarify these steps.

1. Choose a vector and call it x_0. Set $i = 0$.
2. Multiply to get the next approximation for x using the formula $x_{i+1} = Ax_i$.
3. Divide every term in x_{i+1} by the last element of the vector and call the new vector x'_{i+1}.
4. Repeat steps 2 and 3 until x'_i and x'_{i+1} agree to the desired number of digits.

5. The vector obtained in step 4 is an approximate eigenvector corresponding to the dominant eigenvalue. We will call it x.

6. An approximation to the dominant eigenvalue is $x^T Ax$ .This is called the Rayleigh quotient of x.

Remark: Because any constant multiple of an eigenvector is an eigenvector, we did not have to divide by the last element in the vector in step 3. We could have divided by any element or not divided at all. We divided so that our vector would not grow too large and we could tell when we had converged. We divided by the last element of the vector so that we would have a well-defined algorithm for using the power method. The choice of the last element over any of the others was arbitrary. Therefore, if the last element is zero, divide by another element of the vector for that entire problem. When people programme the power method on a computer, they usually divide by,

$$\|x\|_2 = \sqrt{x_1^2 + x_2^2 + \ldots + x_n^2},$$

which is defined as the length of the vector, so that they don't have to worry about whether or not an element is zero.

Remark: Some calculators will not let you divide a vector by a constant. On those calculators, you can multiply by the multiplicative inverse (reciprocal) of the constant.

Remark: You will probably not be able to directly input the Rayleigh quotient into your calculator. It will consider the numerator and denominator as 1 by 1 matrices. We consider 1 by 1 matrices to be the same as real numbers, but your calculator may not consider them the same. Since you cannot divide matrices, your calculator will probably give you an error message.

You have seen the steps to the power method. Let's demonstrate those steps on the matrix .

$$A = \begin{bmatrix} 3 & 6 \\ 1 & 4 \end{bmatrix}$$

For step 1, we arbitrarily chose $x_0 = \begin{bmatrix} 1 \\ 1 \end{bmatrix}$ Let's make a chart for steps 2 and 3.

Sample Theory

The variation that we find among the scores in a sample is not just considered error. In fact, it is thought to represent a specific kind of error: *random error*. When we select a sample at random from a population, we expect that the members of that sample will not all have identical scores on our variable of interest (*e.g.*, test scores). That is, we expect that there will be some variability in the scores of the members of the sample. That's just what happens when you select members of a sample randomly from a population. Therefore, the variation in scores that we see among the members of our sample is just considered random error. The question that we can address using ANOVA is this: Is the average amount of difference, or variation, between the scores of members *of different* samples large or small compared to the average amount of variation *within* each sample, otherwise known as random error (a.k.a. error)? To answer this question, we have to determine three things.

First, we have to calculate the average amount of variation within each of our samples. This is called the mean square within ($MS_{W)}$ or the mean square error ($MS_{e)}$. Second, we have to find the average amount of variation *between* the groups. This is called the mean square between ($MS_{b)}$. Once we've found these two statistics, we must find their ratio by dividing the mean square between by the mean square error. This ratio provides our F value, and when we have our F value we can look at our family of F distributions to see if the differences between the groups are statistically significant. Recall that the formula for calculating an independent t test also involves finding a ratio.

The portion of the fraction is the difference between two sample means, which is analogous to the mean square between *(MSb)* just presented.

The only differences between the two are:

1. Rather than finding a simple difference between two means as in a t test, in ANOVA we are finding the *average* difference between means, because we often are comparing more than two means;
2. We are using the squared value of the difference between the means. The bottom portion of the fraction for the t test is the standard *error* of the difference between two sample means. This is exactly the same as the *average,* or standard, error within groups. In the formula used to calculate the F value in ANOVA, we must square this average within-group error. So, just as in the t test, in ANOVA we are trying to find the

average difference *between* group means relative to the average amount of variation *within* each group.

To find the MS_e and MS_b, we must begin by finding the sum of squares error ($SS_{e)}$ and the sum of squares between ($SS_{b)}$. This sum of squares idea is not new. It is the same old sum of squares introduced back when talking about variance and standard deviation. Sum of squares is actually short for *sum of squared deviations.* In the case of ANOVA, we have two types of deviations. The first is deviations between each score in a sample and the mean for that sample (*i.e.*, error). The second type of deviation is between each sample mean and the mean for all of the groups combined, called the grand mean (*i.e.*, between groups).

To find the sum of squares error:

1. *Subtract the group mean from each individual score in each group*: $(X–X)$.
2. *Square each of these deviation scores*: $(X–X)^2$.
3. *Add them all up for each group*: $\Sigma(X–X)^2$.
4. *Then add up all of the sums of squares for all of the groups*:

The subscripts indicate the individual groups, through the last group, which is indicated with the subscript k.The method used to calculate the sum of squares between groups *(SSb)* is just slightly more complicated than the SS_e formula. To find the SS_b we

1. *Subtract the grand mean from the group mean*: $(X–X_{T)}$; T indicates total, or the mean for the total group.
2. *Square each of these deviation scores*: $(X–X_{T)}^2$.
3. *Multiply each squared deviation by the number of cases in the group*: $[n(X–X_{T)}^2]$.
4. *Add these squared deviations from each group together*: $\Sigma[n(X–X_{T)}^2]$.
5. In the SS_e we subtract the group mean from the individual scores in each group whereas in the SS_b we subtract the grand mean from each group mean.
6. In the SS_b we multiply each squared deviation by the number of cases in each group. We must do this to get an approximate deviation between the group mean and the grand mean for *each case in every group.*

If we were to add the SS_e to the SS_b, the resulting sum would be called the sum of squares total ($SS_{T)}$. A brief word about the

SS_T is in order. Suppose that we have three randomly selected samples of children. One is a sample of 5th graders, another is a sample of 8th graders, and the third is a sample of 11th graders. If we were to give each student in each sample a spelling test, we could add up the scores for all of the children in the three samples combined and divide by the total number of scores to produce one average score.

Because we have combined the scores from all three samples, this overall average score would be called the grand mean, or total mean, which would have the symbol X_T. Using this grand mean, we could calculate a squared deviation score for each child in all three of our samples combined using the familiar formula $(X–X_T)^2$. The interesting thing about these squared deviations is that, for each child, the difference between each child's score and the grand mean is the sum of that child's deviation from the mean of his or her own group plus the deviation of that group mean from the grand mean. So, suppose Jimmy is in the fifth-grade sample. Jimmy gets a score of 25 on the spelling test. The average score for the fifth-grade sample is 30, and the average score for all of the samples combined (*i.e.*, the grand mean) is 35.

The difference between Jimmy's score (25) and the grand mean (35) is just the difference between Jimmy's score and the mean for his group (25–30=–5) plus the difference between his group's mean and the grand mean (30–35=–5). Jimmy's deviation from the grand mean is –10. If we square that deviation score, we end up with a squared deviation of 100 for Jimmy.

Now, if we calculated a deviation score for each child in all three samples and added up all of these deviation scores using the old E $(X–X_T)^2$ formula, the result would be the sum of squares total, or the SS_T. The interesting thing about this SS_T is that it is really just the sum of the SS_b and the SS_e. $SS_T = SSb + SS_e$. This makes sense, because, as we saw with Jimmy, the difference between any individual score and the grand mean is just the sum of the difference between the individual score and the mean of the group that the individual is from plus the difference between that group mean and the grand mean.

Group

Once we have calculated the SS_b and the SS_e, we have to convert them to average squared deviation scores, or MSb and MS_e. This is necessary because there are far more deviation scores in the SS_e than there are in the SS_b, so the sums of squares can be a bit misleading. What we want to know in an ANOVA is whether the *average* difference between the group means is large or small relative to the *average* difference between the individual scores and their respective group means, or the average amount of error within each group. To convert these sums of squares into mean squares, we must divide the sums of squares by their appropriate degrees of freedom.

For the SS_b, remember that we are only making comparisons between each of the groups. The degrees of freedom for the SS_b is always the number of groups minus 1. If we use *K* to represent the number of groups, and *df* to represent degrees of freedom, then the formula for the between groups degrees of freedom is *df=K*–1. So, to convert an *SSb* to an *MSb, we* divide SS_b *by K*- 1. The degrees of freedom for SS_e is found by taking the number of scores in each group and subtracting 1 from each group. So, if we have three groups, our *df for* SS_e will be $(n_{1}–1) + (n_{2}–1) + (n_{3}–1)$. Notice that this is the same formula for the degrees of freedom that was used for the independent samples *t* test. The only difference is that we have one more group here. A simpler way to write this *df* formula is *N-K,* where N is the total number of cases for *all* groups combined and *K* is the number of groups. Once we have this *df,* we can convert the SS_e into an MS_e by simply dividing SS_e *by N–K.* Table contains a summary of the formulas for converting the sums of squares into mean squares.

Once we have found our *MSb and* our MS_e, all we have to do is divide *MSb by* MS_e to find our *F* value. Once we've found our *F* value, we need to look in our table of *F* values (Appendix C) to see whether it is statistically significant.

This table of *F* values is similar to the table of *t* values we used, with one important difference. Unlike *t* values, the significance of *F* values depends on both the number of cases in the samples (*i.e.*, the *df for* MS_e) *and* the number of groups being compared (*i.e.*, the *df for* MS_b). This second *df is* critical, because it is what is used to control for the fact that we are comparing more than two groups. Without it, we might as well conduct multiple *t* tests, and this is problematic for the reasons discussed at the beginning of the chapter. In Appendix C, we can find critical values for *F* associated with different alpha levels. If our observed value of *F* is larger than our critical value of *F,* we must conclude that there are statistically significant differences between the group means.

General Principles

The techniques are based on the same general principles as one-way ANOVA. Namely, they all involve the partitioning of the variance of a dependent variable into its component parts (*e.g.*, the part attributable to between-group differences, the part attributable to within-group variance, or error). In addition, these techniques allow us to examine more complex, and often more interesting questions than is allowed by simple one-way ANOVA. These more advanced statistical techniques involve much more complex formulas than those we have seen previously.

Factorial ANOVA is the technique to use when you have one continuous (*i.e.*, interval or ratio scaled) dependent variable and two or more categorical (*i.e.*, nominally scaled) independent variables. For example, suppose we want to know whether boys and girls differ in the amount of television they watch per week, on average. Suppose we also want to know whether children in different regions of the United States (*i.e.*, East, West, North, and South) differ in their average amount of television watched per week. In this example, average amount of television watched per week is our dependent variable, and gender and region of the country are our two independent variables. This is known as a 2 × 4 factorial analysis, because one of our independent variables has two levels (gender) and one has four levels (region). If we were writing about this analysis in an academic paper, we would write "we conducted a 2 (gender) × 4 (region) factorial ANOVA."

Now when we run our factorial ANOVA, we will get three interesting results. First, we will get two main effects, one for our comparison of boys and girls and one for our comparison of children from different regions of the country. These results are similar to the results we would get if we simply were to run two one-way ANOVAs, with one important difference, which we describe in the next section. In addition to these main effects, our factorial ANOVA also produces an interaction effect, or simply an interaction. An interaction is present when the differences between the groups of one independent on the dependent variable vary according to the level of a second independent variable. Interaction effects are also known as moderator effects. We discuss interactions in greater detail in the next section as well. For now, suffice it to say that interaction effects are often very interesting and important pieces of information for social scientists.

Ideal Situation

Just as with one-way ANOVA, when conducting a factorial ANOVA it is important to determine whether there are roughly an equal number of cases in each group and whether the amount of variance within each group is roughly equal (known as homogeneity of variance). The ideal situation in ANOVA is to have roughly equal sample sizes in each group and a roughly equal amount of variation (*e.g.*, the standard deviation) in each group. If the sample sizes are equal but the standard deviations are significantly different, this is not a big problem. But if the sample sizes *and* the standard deviations are quite different in the various groups, there is a problem. This is particularly likely to be a problem in factorial ANOVA because the combination of independent variables often produces a small number of cases in each group.

In our previous example, suppose we have 20 boys and 20 girls in our entire sample. In addition, suppose that we have 20 children from each of the four regions in our sample. To test the main effects, these numbers are acceptable. That is, it is reasonable to compare 20 boys to 20 girls if we want to know whether boys and girls differ in their average amount of television viewing. Similarly, it is reasonable to compare 20 children from each of the four different regions of the country. But suppose that in the West, our sample of 20 children includes only 5 girls and 15 boys, whereas in the North our sample includes 15 girls and only 5 boys. When we divide up our sample by *two* independent variables, it is easy to wind up with cell sizes that are too small to conduct meaningful ANOVAs. A "cell" is a subset of cases representing one unique point of intersection between the independent variables. In the aforementioned example, there would be eight cells: girls from the West, boys from the West, girls from the South, boys from the South, and so on. When you consider that factorial ANOVAs can have more than two independent variables, the sample can be subdivided a number of times. Without a large initial sample, it is easy to wind up with cells that contain too few cases. As a general rule, cells that have fewer than 10 cases are too small to include in ANOVAs, and cell sizes of at least 20 are preferred.

Factorials

When dividing up the variance of a dependent variable, such as hours of television watched per week, into its component parts, there are a number of components that we can examine. In this section, we examine three of these components: The main effects, interaction effects, and simple effects. In addition, we also present an introduction to the idea of partial and controlled effects.

Occurance

These main effects will each have their own *F* value, and are very similar to the results that would be produced if you just conducted a one-way ANOVA for each independent variable on the dependent variable. However, there is one glorious benefit of looking at the main effects in a factorial ANOVA rather than separate one-way ANOVAs: When looking at the main effects from a factorial ANOVA, it is possible to test whether there are significant differences between the groups of one independent variable on the dependent variable while controlling for, or partialing out the effects of the other independent variables on the dependent variable. Let me clarify this confusing sentence by returning to our example of television viewing.

Suppose that when we examine whether boys and girls differ in the average amount of television they watch per week, we find that there is a significant difference: Boys watch significantly more television than girls. In addition, suppose that children in the North watch, on average, more television than children in the South. Now, suppose that, in our sample of children from the Northern region of the country, there are twice as many boys as girls, whereas in our sample from the South there are twice as many girls as boys. Now I've got a potential problem.

How do we know whether our finding that children in the North watch more television than children in the South is not just some *artifact* caused by the greater proportion of boys in our Northern sample? By "artifact" we mean that the North-South difference is merely a *by-product* of the difference between boys and girls; region of the country is not an important factor in and of itself.

Think about it: If we already know that boys watch more television, on average, than girls, then we would *expect* our Northern sample to watch more television than our Southern sample because there is a greater proportion of boys in our Northern sample than in the Southern sample. So our question is this: How can we determine whether there is a difference in the average amount of television watched by children in the North and South *beyond* the difference caused by the unequal proportions of boys and girls in the samples from these two regions. Phrased another way, is there an effect of region on television viewing *beyond,* or *in addition to* the effect of gender? To answer this intriguing question, we must examine the main effect of region on television viewing after *controlling for,* or *partialing out* the effect of gender. We can do this in a factorial ANOVA. To understand how this is accomplished, keep in mind that what we are trying to do with an ANOVA is to *explain* the variance in our dependent variable (amount of television children watch per week) by dividing that variance up into its component parts. If boys and girls differ in how much they watch television, then part of the variance is explained, or accounted for, by gender.

In other words, we can understand a bit of the differences among children in their weekly television viewing if we know their gender. Now, once we remove that portion of the total variance that is explained by gender, we can test whether any *additional* part of the variance can be explained by knowing what region of the country children are from. If children from the North and South *still* differ in the amount of television they watch, after *partialing out* or *controlling for* the chunk of variance explained by gender, then we know that there is a main effect of region *independent* o/the effect of gender. In statistical jargon, we would say "There is a main effect of region on amount of television watched *after controlling for the effect of gender."* This is powerful information. In factorial ANOVA, it is possible to examine each main effect and each interaction effect when controlling for *all* other effects in the analysis.

A second benefit of factorial ANOVA is that it allows researchers to test whether there are any statistical interactions present. Interactions can be a complex concept to grasp. Making the whole issue even more confusing is that the level of possible interactions increases as the number of variables increases.

For example, when there are two independent variables in the analysis, there are two possible main effects and one possible two-way interaction effect (*i.e.*, the interaction between the two independent variables). If there are three independent variables in the analysis, there are three possible main effects, three possible two-way interaction effects, and one possible three-way interaction effect. The whole analysis can get very complicated very quickly. To keep things simple, let's take a look at two-way interactions first.

In our television-viewing example, suppose that we randomly selected 25 boys and 25 girls from each of the four regions of the country, measured the number of hours each child spent watching television, and calculated the averages for each group. (Note: Unlike the example provided earlier, in this example there are equal numbers of boys and girls from each region in the sample.)

The overall averages by gender presented in the last column indicate that there appears to be a main effect for gender, with boys watching an average of 22.5 hours of television per week and girls watching an average of only 15 hours per week. When we look at the overall averages presented for each region (bottom row), we can see that children in the North watch more television, on average, than do children in the other three regions. Therefore, we can tell that there appear to be main effects for gender and region on amount of television watched. Notice that it is said *"appear* to be main effects." To determine whether these main effects of statistically significant, we would have to determine the probability of obtaining differences of this size between randomly selected groups of this size.

Once we have examined the main effects, we can turn our attention to the possible interaction effects. To do this, we need to examine the means in each of the eight cells presented in Table (*i.e.*, Northern boys, Northern girls, Eastern boys, Eastern girls, etc.). When we examine these means, we can see that in the North, East, and West, boys watch an average of 5 hours more television per week than girls.

But in the South, boys watch an average of 15 more hours of television than girls. Therefore, it appears that the differences in the amount of television watched by girls and boys are not uniform across the four regions of the country. In other words, the relationship between gender and amount of television watched *depends on,* or *is moderated by,* the region of the country. Because the definition of a two-way interaction is that the relationship between an independent variable and a dependent variable is moderated by a second independent variable, we appear to have a two-way interaction here.

We can also see evidence of a main effect for region, although this effect is somewhat less clear than the main effect for boys. We see the region effect by noting that for both boys and girls, the average amount of television viewing is higher in the North than in either the East or West regions. This main effect is complicated a bit, however, by the presence of the interaction.

Notice that whereas the mean is lower in the South than in the North for girls (supporting our main effect for region), the mean for boys in the South is equal to the mean for boys in the North. This raises a difficult question: When we say there is a main effect for region, with children in the North watching more television, on average, than children in the other three regions, are we being accurate?

Interaction Effect

Some argue that it makes little sense to interpret main effects at all when there is an interaction effect present, because the interaction effect essentially modifies the meaning of the main effect. In the preceding example, the main effect for region that shows children in the North watch more television than children elsewhere is really only true within the girls sample.

In fact, boys in the South watch as much television as boys in the North, and girls in the North do not watch more television than boys in any region. Therefore, some would argue that we should just describe the nature of the interaction, and not interpret the main effects. The logic of this argument is as follows: If we say that children in the North watch more television than children in other regions, the statement is misleading because it is not true for boys. To be accurate, we should just say that *girls* in the North watch more television that *girls* in other regions.

Others, ourself included, think it makes sense to interpret all of the effects and to consider them in relation to each other. Returning to our earlier example, we can see that there is a main effect for gender, with boys watching more television, on average, than girls. We can also see that this effect is especially pronounced in the South. In addition, we can say that overall, when we *combine* the samples of boys and girls together, there is a main effect for region such that Northern children watch more television than children in other regions, on average.

When we add the consideration of the interaction effect, we can further argue that this overall effect is due *primarily* to differences within the sample of girls, and less to variation within the sample of boys. It is possible to get an interaction effect without a main effect. In this example, boys and girls have equal means, as do children in each of the four geographic regions). Therefore, it makes sense to report and interpret significant main effects, even in the presence of an interaction effect. The key is to provide enough information so that readers of your results can makes sense of them.

Suppose that we were to examine the math skills of boys and girls in two different types of mathematics programmes. Students in the "Traditional" programme study math in the usual way, reading a textbook and working out math problems in class. Students in the

"Experimental" programme work in groups to solve problems collaboratively and work with more real-world, applied problems. After one year, we give a math test to 25 randomly selected boys and 25 randomly selected girls from each math programme.

Methods

Once we have found our main and interaction effects in factorial ANOVA, we can conduct one final set of analyses to examine the simple effects. The methods used to calculate the simple effects and determine whether they are statistically significant are analogous to the post-hoc tests described in Chapter. What simple effects analysis allows us to do is to test whether there are significant differences in the average scores of any particular cells. One of the benefits of simple effects analysis is that is allows us to better understand some of the complexities in our data, particularly how to make sense of significant interaction effects.

Returning to our sample data presented in Figure we can see that we have four cells: girls in the Traditional math programme, Traditional boys, Experimental girls, and Experimental boys. With a simple effects analysis, we can test whether boys in the Traditional math programme ($X = 50$) had significantly higher average math test scores than did boys in the Experimental programme (X " 35). We could also test whether boys and girls in the Traditional programme differed significantly. Perhaps most important for helping us understand the interaction effect, we can test whether girls in the Experimental programme had higher average math test scores than students in each of the three other groups.

Analysis

Earlier in this chapter, we suggested that one of the benefits of conducting factorial ANOVAs is that it allows us to determine whether groups differ on some dependent variable while controlling for, or partialing out, the effects of other independent variables. A closely related concept that applies to all types of ANOVA, including one-way, factorial, and repeated-measures, is the use of covariates in these analyses. In analysis of covariance (ANCOVA), the idea is to test whether there are differences between groups on a dependent variable after controlling for the effects of a different variable, or set of variables. The difference between an ANCOVA and the types of controlled variance we described earlier is that with an ANCOVA, the variable(s) that we are controlling, or partialing out, the effects of is not necessarily an independent variable. Let me explain.

In our earlier example, we was able to test whether boys and girls differed in the amount of television they watched while controlling for the effects of which region of the country they lived in (the second independent variable), as well as the interaction between the two independent variables.

But in an ANCOVA analysis, we can control the effects of variables *besides* independent variables. For example, we could use socioeconomic status (SES) as a covariate and test whether children in different regions of the country differ in the amount of television they watch *after* partialing out the effects of their SES. Suppose that our sample of children from the North is less wealthy than our samples from the three other regions. Suppose further than children from poorer families tend to watch more television than children from wealthier families. Because of this, our earlier results that found greater television watching among children in the Northern region may simply be due to the fact that these children are less wealthy than children in the other regions. With ANCOVA, we can test whether the difference in the viewing habits of children from different regions is due strictly to differences in SES, or whether there are regional differences *independent* of the effects of SES. This is particularly handy because even though factorial ANOVA only allows us to use categorical (*i.e.*, nominally scaled) independent variables, with ANCOVA we can also control the effects of continuous (*i.e.*, intervally scaled) variables.

Notation Factorial

We wanted to see whether high school boys and girls differed in their self-efficacy (*i.e.*, a main effect for gender), whether students with relatively high grade point averages (GPA) differed from those with relatively low GPAs in their self-efficacy (*i.e.*, a main effect for GPA), and whether there was an interaction between gender and GPA on self-efficacy. Gender, of course, is a two-category independent variable. To make GPA a two-category variable, we divided students into high- and low-GPA groups by splitting the sample in two using the median GPA. This allowed me to perform a 2 (gender) × 2 (GPA) factorial ANOVA. Self-efficacy was measured using a survey with a 5-point scale (1="not at all confident" and 5="very confident"). Our sample consisted of 468 high school students.

The results presented in Table begin with descriptive statistics. These statistics are presented separately by subgroups (*e.g.*, low-achieving girls, high-achieving girls, all girls combined, low-achieving boys, etc.). The means and standard deviations presented are for the dependent variable, self-efficacy. By glancing over the means we can see that the boys in our sample had slightly higher average feelings of self-efficacy than did the girls, and this difference appears to be largest among the boys and girls in the low-GPA group.

Turning our attention to the ANOVA results, there are a number of important features to notice. In the far left column titled "Source, " there are the various *sources* of variation in self-efficacy. These are the different ways that the variance of the dependent variable, self-efficacy, is sliced up by the independent variables. The first source is called the "Corrected Model." This is the combination of all of the main and interaction effects. If covariates were used, these effects would be included in the "Corrected Model" statistics. Reading from left to right, we can see that the full model has a sum of squares (11.402), which when divided by three degrees of freedom *("df")* produces a "Mean Square" of 3.801. When we divide this by the mean square error a few rows down (MS_e =.649) we get an F value of 5.854. This has a "Sig." of.001 (in other words, p =.001). Because this value is less than.05, the overall model is statistically significant. But is it *practically* significant? In other words, gender, GPA level, and the interaction of these two *combined* only explain 3.6 Per cent of the variance. Although this is *statistically* significant, this may not be a big enough effect size to be considered *practically* significant. Remember that statistical significance is influenced by sample size, and 468 cases is a pretty large sample. An effect size of.036, in contrast, is not affected by sample size and therefore may be a better indicator of practical significance.

In addition to the F value, p value, and effect size for the entire model, SPSS prints out statistics for each of the main effects and the interaction effect as well. Here we can see that gender is a statistically significant predictor of self-efficacy ("Sig."=.023), but the effect size of.011 is very small. By looking at the overall means for girls and boys in the top portion of Table, we can see that boys (X_{boys} = 4.0031) have the slightly higher average feelings of self-efficacy in English than girls (*Xgirls* = 3.8438). GPA has a larger F value (F = 10.646) and is statistically significant, but it also has a small eta squared value (.022). Students in the high-GPA group had slightly higher average feelings of self-efficacy (X_{high} = 4.0381) than did students in the low-GPA group (X_{low} = 3.793). The gender by GPA interaction was not statistically significant and has a tiny effect size. Overall, then, the statistics presented in Table reveal that although there are statistically significant main effects for gender and GPA on self-efficacy, and the overall model is statistically significant, these effect sizes are quite small and suggest that there is not a strong association between either gender nor GPA with self-efficacy among this sample of high school students.

There are two other features of the SPSS output presented in Table worth noting. First, the sum of squares that SPSS uses by default in a factorial ANOVA is called "Type III" sum of squares. This means that when SPSS calculates the sum of squares for a particular effect, it does so by accounting for the other effects in the model. So when the sum of squares for the gender effect is calculated, for example, the effect of GPA and the gender by GPA interaction effects have been partialed out already. This allows us to determine the *unique* effect of each main effect and interaction effect. Second, notice that the F value for each effect is obtained by dividing the mean square for that effect by the mean square error.

Single Measures

One type of study in which a paired t test would be used is when we have two scores for a single group on a single measure. For example, if we had a group of third-grade students and we gave them a test of their math abilities at the beginning of the school year and again at the end of the school year, we would have one group (third graders) with two scores on one measure (the math test). In this situation, we could also use a repeated-measures analysis of variance (ANOVA) to test whether students scores on the math test were different at the beginning and end of the academic year.

Repeated-measures ANOVA has a number of advantages over paired t tests, however. First, with repeated-measures ANOVA, we can examine differences on a dependent variable that has been measured at more than two time points, whereas with an independent t test we can only compare scores on a dependent variable from two time points. Second, as discussed on factorial ANOVA, with a repeated-measures ANOVA we can control the for effects of one or more covariates, thereby conducting a repeated-measures analysis of covariance (ANCOVA).

Techniques

The most basic form of a repeated-measures ANOVA occurs when there is a single group (*e.g.*, third graders) with two scores (*e.g.*, beginning of the year, end of the year) on a single dependent variable (*e.g.*, a mathematics test). This is a very common model that is often used in simple laboratory experiments. For example, suppose we wanted to know whether drinking alcohol affects the reaction time of adults when driving. We could take a group of 50 adults and test their reaction time when driving by flashing a red light at each one of them when they are driving and measuring how long it takes for each one to apply the brakes. After calculating the average amount of time it takes this group to apply the brakes when sobre, we could then ask each member of our group to consume two alcoholic drinks and then again test their reaction time when driving, using the same methods. In this example, I've got one group (50 adults) with two scores on one dependent variable (reaction time when driving).

After the second measure of reaction time, we could ask each of our participants to consume two more alcoholic drinks and again test their reaction time. Now we've got three measures of reaction time that we can use in our repeated-measures ANOVA. Notice that our dependent variable is always the same measure (reaction time), and our group is always the same (sample of 50 adults).

The results of our repeated-measures ANOVA will tell me whether, on average, there are differences in reaction time across our three trials. If there are, we might logically conclude that drinking alcohol affects reaction time, although there may be other explanations for our results (*e.g.*, our participants may be getting tired or bored with the experiment, they may be getting used to the test situation, etc.).

In a slightly more advanced form of the test of reaction time, we could include a covariate. In the previous example, suppose that we found the reaction time was fastest when our participants were sobre, a bit slower after two drinks, and a lot slower after four drinks. Suppose that we publish these results and the national beer, wine, and liquor companies become worried that, because of our study, people will stop drinking their products for fear of getting in automobile accidents. These producers of alcoholic drinks begin to criticize our study. They suggest that because equal amounts of alcohol generally have greater effects on those who weigh less than on heavier people, our results may have been skewed by the effects of alcohol on the lighter people in our study. "Although the effects of two alcoholic drinks may impair the reaction time of lighter people, even four alcoholic drinks will not impair the reaction time of heavier people" said the United Alcohol Makers of America (a fictitious group). Stung by the criticism of the UAMA, we decide to replicate our study, but this time we use weight as a covariate. Again, we measure participants' reaction time when driving completely sobre, after two alcoholic drinks, and after four alcoholic drinks. In addition, this time we weigh each of our participants.

Now when we analyse our data, we include our weight covariate. We find that, after controlling, or partialing out the effects of weight, there is no difference in the reaction time of participants before they have any drinks and after they have two drinks, but after four drinks our participants react more slowly, on average, than they did after zero or two drinks. These results suggest that drinking may increase the reaction time of lighter people after only two drinks, but it seems to impair the reaction time of people, regardless of weight, after four drinks.

Still bothered by our results, the UAMA suggests that our results are skewed because we did not look at the effects of drinking on reaction time separately for men and women. "Women are more dramatically affected by alcohol than men, regardless of weight" claims the UAMA. They argue that although consuming four alcoholic drinks may slow the reaction time of women, it will not have an effect on heavy men. Though we are dubious of the argument that heavy men should have their rights to drink and drive protected, in the name of science we decide to conduct one final study.

In this study, again with 50 adults (25 women and 25 men) of various weights, we again test their reaction time while driving after zero, two, and four alcoholic drinks. Now I've got one dependent variable (reaction time) measured at three time points, one covariate (weight), and one independent group variable (gender of participant). Notice that although number of drinks is technically an independent variable, it is not a categorical, or group, variable. In other words, we do not have three independent groups (the zero-drink group, the two-drink group, and the four-drink group). Rather, we have three *dependent,* or repeated, measures of the same dependent variable, reaction time. When we examine the results of our study, we find that, after controlling for the effects of our covariate (weight), there is still no difference in reaction time measured after zero and two drinks, but still slower reaction time, on average, after four drinks. In addition, we find no interaction between gender and number of drinks on reaction time. This tells us that *both* men and women have slower reaction time after four drinks, regardless of their weight.

To summarize, our three different repeated-measures ANOVAs produced the following results. The first one found that adults' reaction times while driving were slower, on average, after two drinks and slower still after four drinks. Our second test included the covariate of weight, and we found that when we control for the effects of weight, reaction time is not slower after two drinks but is slower after four drinks. Finally, in our third analysis, we examined whether changes in reaction time after two and four drinks, when controlling for weight, was different for men and women. found that it was not. These three analysis provide a snapshot of how repeated-measures ANOVA works and what information it can provide. Now let's take a closer look at how it works.

Complicated Form

A slightly more complicated form of repeated-measures ANOVA can be produced by adding one or more covariates to the model. As discussed earlier in this chapter, as well as in factorial ANOVA, covariates can be used to partial out some portion of the variance in the dependent variable.

One could argue that the results of our repeated-measures ANOVA were skewed by the scores of the more intelligent students in our sample. Although the students in our sample scored higher on the vocabulary test at the end of fourth grade than they did at the end of third grade, we must keep in mind that the change in scores over time represents an *average* change. Some students in our sample improved quite a bit over time, whereas others did not increase at all, and one (Subject 5) actually declined. So it is possible that this overall average improvement over time was caused by large increases among the brightest students. To explore this hypothesis, we conduct a new repeated-measures ANOVA, but this time we include a covariate: IQ test scores.

When we conduct our repeated-measures ANCOVA, we now have three ways of partitioning the total variance in our vocabulary test scores. First, there is the portion of variance that is accounted for by our covariate, IQ test scores. If students' IQ test scores are related to (*i.e.*, correlated with) their vocabulary test scores, then the IQ test scores will explain, or account for, some percentage of the variance in students' vocabulary test scores. Second, after partialing out the portion of variance attributable to IQ test scores, we can see whether any of the remaining variance in vocabulary test scores is accounted for by *changes* in vocabulary test scores from third to fourth grade.

In other words, once we control for the effects of IQ test scores, do the scores of our sample change significantly from Time 1 (third grade) to Time 2 (fourth grade), on average? Is there still a *within-subjects* effect after controlling for IQ test scores? Finally, after accounting for the variance in vocabulary test scores that is attributable to the covariate (*i.e.*, IQ test scores) and the within-subjects effect (*i.e.*, changes from third to fourth grade), there will still be some variance in vocabulary test scores that is not explained. This is error variance, which is the same as the random variance that we normally find between different members of the same sample.

To reiterate, when a covariate (or several covariates) are added to the repeated-measures ANOVA model, they are simply included to "soak up" a portion of the variance in the dependent variable. Then, we can see whether there are any within-subject differences in the scores on the dependent variable, when *controlling for, or partialing out* that portion of the variance accounted for by the covariate(s). In the example we have been using, the addition of the IQ score covariate allows us to answer this question: Do students' vocabulary test scores change, on average, from third to fourth grade *independently* of their IQ scores? Phrased another way, we can ask whether, when controlling for IQ, students' vocabulary test scores change from third to fourth grade.

Probability

History

A gambler's dispute in 1654 led to the creation of a mathematical theory of probability by two famous French mathematicians, Blaise Pascal and Pierre de Fermat. Antoine Gombaud, Chevalier de Méré, a French nobleman with an interest in gaming and gambling questions, called Pascal's attention to an apparent contradiction concerning a popular dice game. The game consisted in throwing a pair of dice 24 times; the problem was to decide whether or not to bet even money on the occurrence of at least one "double six" during the 24 throws. A seemingly well-established gambling rule led de Méré to believe that betting on a double six in 24 throws would be profitable, but his own calculations indicated just the opposite.

This problem and others posed by de Méré led to an exchange of letters between Pascal and Fermat in which the fundamental principles of probability theory were formulated for the first time. Although a few special problems on games of chance had been solved by some Italian mathematicians in the 15th and 16th centuries, no general theory was developed before this famous correspondence.

The Dutch scientist Christian Huygens, a teacher of Leibniz, learned of this correspondence and shortly thereafter published the first book on probability; entitled *De Ratiociniis in Ludo Aleae*, it was a treatise on problems associated with gambling. Because of the inherent appeal of games of chance, probability theory soon became popular, and the subject developed rapidly during the 18th century. The major contributors during this period were Jakob Bernoulli and Abraham de Moivre. In 1812 Pierre de Laplace introduced a host of new ideas and mathematical techniques in his book, *Théorie Analytique des Probabilités*. Before Laplace, probability theory was solely concerned with developing a mathematical analysis of games of chance. Laplace applied probabilistic ideas to many scientific and practical problems. The theory of errors, actuarial mathematics, and statistical mechanics are examples of some of the important applications of probability theory developed in the 19th century.

Like so many other branches of mathematics, the development of probability theory has been stimulated by the variety of its applications. Conversely, each advance in the theory has enlarged the scope of its influence. Mathematical statistics is one important branch of applied probability; other applications occur in such widely different fields as genetics, psychology, economics, and engineering. Many workers have contributed to the theory since Laplace's time; among the most important are Chebyshev, Markov, von Mises, and Kolmogorov.

One of the difficulties in developing a mathematical theory of probability has been to arrive at a definition of probability that is precise enough for use in mathematics, yet comprehensive enough to be applicable to a wide range of phenomena. The search for a widely acceptable definition took nearly three centuries and was marked by much controversy. The matter was finally resolved in the 20th century by treating probability theory on an axiomatic basis. In 1933 a monograph by a Russian mathematician A. Kolmogorov outlined an axiomatic approach that forms the basis for the modern theory. Since then the ideas have been refined somewhat and probability theory is now part of a more general discipline known as measure theory."

The term Probability, as the name suggests, refers to a calculation, a mathematical estimate concerning the possibility of a particular outcome in a situation where there are a number of different possible outcomes. If, for example, we know that in a certain situation (*e.g.*, a roll of the dice) there is a determined number of possible outcomes (*e.g.*, six possible faces of the die), we can calculate the probability of a particular outcome or of a particular sequence of outcomes if the event is repeated. For example, with a normal die in a fair roll, there is a 1 in 6 chance of any particular face of the die showing on the top (since there are six surfaces on the cube); in a fair toss of a normal coin there is a 1 in 2 chance of a head appearing on top, a 1 in 2 chance of a tail appearing on top, and a 2 in 2 chance of a head or a tail appearing on top. It is important to stress at the outset that probability is a *theoretical* calculation of the likelihood of a particular outcome. Thus, it is based, not upon observation of actual events (*e.g.*, listing the results from several tosses of the coin) but on calculations which are based on certain assumptions (*e.g.*, that the coin or die toss is fair, that there are no hidden factors, like cheating, influencing the outcome).

Because the probability of an outcome is a theoretical calculation, knowing the probability does not mean that we can always predict a particular outcome with certainty (please understand this very important point). Often the result of a particular event or series of events will not match the theoretical probabilities. For example, in a sequence of ten coin tosses, theory tells us that the most probable result will be 5 outcomes with heads on top and 5 outcomes with tails on top. But that calculation does not mean that, if we observe an actual 10-toss sequence, we will always get that most probable result.

Checking probabilities (that is, the theoretical calculations) against the observed outcomes of a real event is, as we shall see, an essential part of the study of probability and statistics (and a central exercise in this module). But always keep in mind this opening point: the probability of an event is a theoretical calculation; the observed frequency with which an event actually occurs is something different: the first is based upon mathematical analysis of an event (independent of observation); the second is based upon an actual experiment. Probability theory has a number of extremely important uses. A simple but powerful application is its ability to assist us in establishing the mathematical likelihood of a particular outcome and thus to advise ourselves of the prudence of a certain action (*e.g.*, betting). For instance, to take a very simple example, in a fair toss of a coin, theory indicates that the probability of the coin landing with heads on top is 50 per cent, with tails on top 50 per cent, and with either heads of tails on top 100 per cent. In a bet on the result of a coin toss, therefore, one would be unwise to wager, say, ten dollars that a head would appear against your opponent's bet of two dollars that a tail would appear.

Probability of an Event

We can express the probability of an event mathematically in a number of ways, but the most common is as a decimal between 1 and 0. Thus, a probability of 1 ($p = 1$) is certainty that the event will occur. A probability of 0 ($p = 0$) is certainty that event will not occur. Thus, all probabilities for an event are expressed as a decimal between 0 and 1. The probability that a single fair coin toss will produce a head on top is expressed as follows: $p = .5$. This we might also express as meaning that the probability is 5 out of 10 or 50 per cent. The decimal notation is the most common and will be standard throughout the rest of this workbook.

In the same way, if $p = .75$, that means that the outcome will theoretically occur 75 times out of 100; if $p = .234$, that means that the outcome will theoretically occur 234 times out of 1000; if $p = .01345$, that means the outcome will theoretically occur in 1345 cases out of 100,000.

It will be clear that any value for p greater than.5 means that the result is theoretically more likely to occur than not; a value for p less than.5 means that the result is more likely not to occur than it is to occur. The smaller the value of p, the less likely that outcome.

Probability Outcome

One of the major applications of probability is to check the theoretical possibility of a particular outcome (calculated mathematically) with the observed result of an event. If we find that the observed frequency of a particular occurrence departs widely from theoretical probabilities (*i.e.*, the result is occurring more frequently than we would expect from our calculations), sometimes we have good reason to investigate the event more carefully.

For example, in a fair toss of a coin, we theoretically calculate p as.5 for a head to appear on any single toss. But what is the probability that someone will throw the coin four times in a row and have the outcome heads each time? We can calculate the probability by multiplying the probabilities together. Since there are four tosses and the probability of a head is.5 on each of these, the probability of a sequence of four heads is as follows:

$$p = .5 \times .5 \times .5 \times .5 = .0625$$

In other words, in a sequence of four consecutive coin tosses, probability theory indicates that in 625 cases out of 10,000 (or 1 outcome in 16) four heads will come up in a row. Thus, if you were making a bet in a number of different trials of four consecutive coin tosses that four heads would appear in a row, you could theoretically expect to win 1 out of every 16 trials and lose 15 out of 16 trials.

Now, this figure of 1 in 16 (p =.0625) does not claim that in every 16 trials of a four-coin-toss sequence, one result will *always* be four heads. The theory states that, on average, in a sequence of four-toss trials, the most likely outcome is that there will be four heads appearing in 1 out of every 16 trials. But in any particular series of four-toss trials, it is quite possible for more than sixteen four-toss trials to happen before the first four-head result appears or for a four-head result to occur before sixteen trials are completed or more than once in one sixteen four-toss sequence. As we shall see, the more times one repeats the four-toss sequence, the closer the observed result will come to the theoretically calculated result.

Thus, given that one four-head sequence every sixteen throws is the most likely (but not the inevitable) outcome of a series of four-toss sequences, if you were betting on the outcome of many four-toss sequences (that there would be four consecutive heads appearing), for the game to be truly fair, you should put up $1 against your opponent's $15 for each four-coin sequence (*i.e.*, if four heads appear you win $15; if they do not your opponent wins $1). With a great many four-toss trials, you and your opponent should come out more or less equal. Hence, if you are betting $1 that four heads will appear in a four-toss sequence of coins and if your opponent is unwilling to put up $15, you should not undertake the bet (the probabilities are against you). If she is willing to put up more than $15, then you should take the bet (the probabilities are in your favour).

When the probability of an event (theoretically calculated) is very low and yet the event occurs, sometimes we are entitled to be suspicious or alarmed at the result. Such moments are an invitation to investigate the event to see if there are hidden factors at work (*e.g.*, loaded dice, cheating, trick coins, some hitherto unknown cause). When do such moments occur? At what point is the probability so unlikely (*i.e.*, so low) that we should investigate further?

Well, let us first, as an example, consider the probability of someone's tossing a coin with ten consecutive heads appearing. Using the same way of calculating the probability as we did above (multiplying together the probabilities for a single event), we can calculate the theoretical probability of ten consecutive heads appearing as follows:

$$p = .5 \times .5 \times .5 \times .5 \times .5 \times .5 \times .5 \times .5 \times .5 \times .5 = .000977$$

This rounded off result indicates that the probability for ten consecutive heads in a series of ten fair coin tosses is 977 out of 1,000,000. Obviously the probability here is much smaller than for the four-head sequence ($p = .0625$), and we would be right to be much more surprised at such a result. But do we have the right to be suspicious? In other words, how improbable does an event have to get before we begin to think that there is something strange going on?

Scientists and social scientists have set conventional standards for how much an observed result has to differ from the theoretical expectation in order for the result to be considered significant (*i.e.*, worthy of notice or of further investigation, because it is not likely to have occurred by chance). That level is $p = .05$ (or 5 results in 100). If there is more than a probability of.05 (*i.e.*, if p is greater than.05) that an event could have occurred by chance, scientists do not usually pay much attention to it. If p is less than.05, however, then the result is considered significant and warrants analysis, because something other than chance may be affecting the outcome.

Thus, for example, someone's tossing a coin and getting four consecutive heads (where $p = .0625$) would not prompt any enquiry; the probability is low, but not so low that we need to start wondering about the fairness of the toss (remember the cut off point is $p = .05$). However, someone's tossing a coin and getting ten heads in a row ($p = .000977$ or, rounded off.001) is significant. The probability is so low (1 chance in 1000) that we have some grounds for thinking the result might not be produced by chance and that it is time to check the coin.

The justification for this convention of a probability of.05 as the dividing line between significant and non-significant results is not mathematical but methodological and practical. Given the constraints, economic for example, upon investigation, it has been found to pay, in terms of success in explaining the world, to concentrate research in areas where such a degree of significance is obtainable. In some tests, as we shall see, where the procedure requires a much higher level of certainty, the cut-off line is sometimes reduced to.01 (or 1 chance in 100). In general, however, the figure of.05 probability is the accepted level below which results become significant and above which we can still ascribe the results to chance.

Note the very important point that a low probability (*e.g.*, p =.001) does not indicate impossibility. It may well be that the ten-head coin-toss sequence, although very very improbable, is fair (*i.e.*, produced by chance). The very low figure for p is a warning that something may be (and probably is) interfering with the result, but the figure does not *prove* such interference. These basic principles indicate one important aspect of probability, its use as a diagnostic tool, a means of recognizing that in a natural event something may be interfering with the results (*e.g.*, cheating or some hidden factor). Significant discrepancies between theoretical calculations of probable outcomes and the observed results of actual outcomes are often the first clue that something needs investigating. And from this initial observation, the investigators often discover something they had not known about before (*e.g.*, a new disease, like Legionnaire's Disease or AIDS, or a new natural phenomenon).

Investigation

The basis upon which the mathematical treatment of probability rests is as follows: we assume that the phenomenon we wish to investigate can be described in terms of a given number of different possible but equally probable outcomes. For instance, in a simple coin toss there are two equally probable outcomes, heads or tails. In one roll of a die, there are six equally probable outcomes, represented by the six sides of the die, each with a number of dots from 1 to 6. In drawing a card from a full and well shuffled normal deck with no jokers, we know that there are four equally probable outcomes for the suit of the drawn card (hearts, diamonds, spades, or clubs), there are two possible outcomes for the colour of the card (red or black), and there are thirteen possible outcomes for the value of the card. Naturally we are assuming that in all these trials the procedure is fair (*i.e.*, no cheating).

Multiple outcomes (*e.g.*, so many heads in consecutive coin tosses or so many sixes in consecutive rolls of a single dice) can be calculated from the probability of a single event (as we did above by multiplying the frequency of repeated events by the probability of the single outcome in one event). Events which cannot be described in terms of equally probable outcomes are beyond the scope of probability theory discussed here.

The probability of any specific outcome is given by the number which is equal to the number of possibilities which involve that outcome divided by the total number of possibilities. For example, in a simple coin toss, there is only one possibility that a head will appear, and the total number of possible outcomes is two. Thus, the probability that a head will appear is 1 divided by 2 or.5. In a roll of a single die, there is one chance that a six will appear, and there are six possible outcomes. Therefore the probability of a six appearing on a single roll of a normal die is 1 divided by 6 (p =.1667). If we draw a card from a normal full deck, there is only one chance that a heart will be on the card, and there are four possibilities for different suits. Therefore the probability of drawing a heart from a full deck is 1 divided by 4 (p =.25). What is the probability that on a single roll of a die, the result will be an even number? Well, there are three possible even outcomes (the numbers 2, 4, and 6) and there are 6 possible outcomes (1, 2, 3, 4, 5, 6). Thus, the probability that on a single roll of a die the result will be an even number of dots is 3 divided by 6: p =.5.

Random Trial or Random Experiment

When the outcome of any experiment can not be predicted precisely then the expeiment is called a random trial or random experiment. In other words, if a random experiment is repeated under identical conditions, the outcome will vary at random as it is impossible to predict about the performance of the experiment. For example, if we toss a honest coin or roll an unbiased dice, we may not get the same results as our expectations.

Sample Space

The totality of all the outcomes or results of a random experiment is denoted by Greek alphabet Ω or English alphabets and is called the sample space. Each outcome or element of this sample space is known as a sample print.

Event

Any subset of a sample space is called an event. A sample space S serves as the universal set for all questions related to an experiment 'S' and an event A w.r.t it is a set of all possible outcomes favorable to the even t A.

For example:

- *A random experiment*:- flipping a coin twice
- *Sample space*: W or $S = \{(HH), (HT), (TH), (TT)\}$
- *The question*: "both the flipps show same face"
- *Therefore, the event A*: $\{(HH), (TT)\}$

Equally Likely Events

All possible results of a random experiment are called equally likely outcomes and we have no reason to expect any one rather than the other. For example, as the result of drawing a card from a well shuffled pck, any card may appear in draw, so that the 52 cards become 52 different events which are equally likely.

Mutually Exclusive Events

Events are called mutually exclusive or disjoint or incompatible if the occurrence of one of them precludes the occurrence of all the others. For example in tossing a coin, there are two mutually exclusive events viz turning up a head and turning up of a tail. Since both these events cannot happen simultaneously. But note that events are compatible if it is possible for them to happen simultaneouly. For instance in rolling of two dice, the cases of the face marked 5 appearing on one dice and face 5 appearing on the other, are compatible.

Exhaustive Events

Events are exhaustive when they include all the possibilities associated with the same trial. In throwing a coin, he turning up of head and of a tail are exhaustive events assuming of course that the coin cannot rest on its edge.

Independent Events

Two events are said to be independent if the occurrence of any event does not affect the occurrence of the other event. For example in tossing of a coin, the events corresponding to the two successive tosses of it are independent. The flip of one penny does not affet in any way the flip of a nickel.

Dependent Events

If the occurrence or non-occurrence of any event affects the happening of the other, then the events are said to be dependent events. For example, in drawing a card from a pack of cards, let the event A be the occurrence of a king in the 1st draw and B be the occurrence of a king in the 1st draw and B be the occurrence of a king in the second draw. If the card drawn at the first trial is not replaced then events A and B are independent events.

Note:

- If an event contains a single simple point i.e. it is a singleton set, then this event is called an elementary or a simple event.
- An event corresponding to the empty set is an "impossible event."

- An event corresponding to the entire sample space is called a 'certain event'.

Complementary Events

Let S be the sample space for an experiment and A be an event in S. Then A is a subset of S. Hence $\overline{A}$, the complement of A in S is also an event in S which contains the outcomes which are not favorable to the occurrence of A i.e. if A occurs, then the outcome of the experiment belongs to A, but if A does not occur, then the outcomes of the experiment belongs to It is obvious that A and are mutually exclusive. A Ç = f and A È = S. If S contains n equally likely, mutually exclusive and exhaustive points and A contains m out of these n points then contains $(n - m)$ sample points.

Definitions of Probability

We shall now consider two definitions of probability:

1. Mathematical or a priori or classical.
2. Statistical or empirical.

Mathematical (or A Priori or Classic) Definition

If there are 'n' exhaustive, mutually exclusive and equally likely cases and m of them are favorable to an event A, the probability of A happening is defined as the ratio m/n.

Expressed as a formula:

Expressed as a formula:

$$P(A) = \frac{m}{n} = \frac{\text{number of favorable outcomes}}{\text{total number of possible outcomes}}$$

$$P(A) = \frac{n(A)}{n(S)}$$

This definition is due to 'Laplace.' Thus probability is a concept which measures numerically the degree of certainty or uncertainty of the occurrence of an event.

For example, the probability of randomly drawing taking from a well-shuffled deck of cards is 4/52. Since 4 is the number of favorable outcomes (*i.e.* 4 kings of diamond, spade, club and heart) and 52 is the number of total outcomes (the number of cards in a deck).

If A is any event of sample space having probability P, then clearly, P is a positive number (expressed as a fraction or usually as a decimal) not greater than unity. 0 £ *P* £ 1 i.e. 0 (no chance or for impossible event) to a high of 1 (certainty). Since the number of cases not favorable to *A* are $(n - m)$, the probability q that event *A* will not happen is, $\frac{n-m}{n}$ q = or q = $1 - m/n$ or $q = 1 - p$. Now note that the probability q is nothing but the probability of the complementary event A i.e.

Thus $p(\overline{A}) = 1 - p$ or $p(\overline{A}) = 1 - p(\overline{A})$

so that $p(A) + p(\overline{A}) = 1$ i.e. $p + q = 1$

Relative Frequency Definition

The classical definition of probability has a disadvantage i.e. the words 'equally likely' are vague. In fact, since these words seem to be synonymous with "equally probable". This definition is circular as it is defining (in terms) of itself. Therefore, the estimated or empirical probability of an event is taken as the relative frequency of the occurrence of the event when the number of observations is very large.

The Laws of Probability

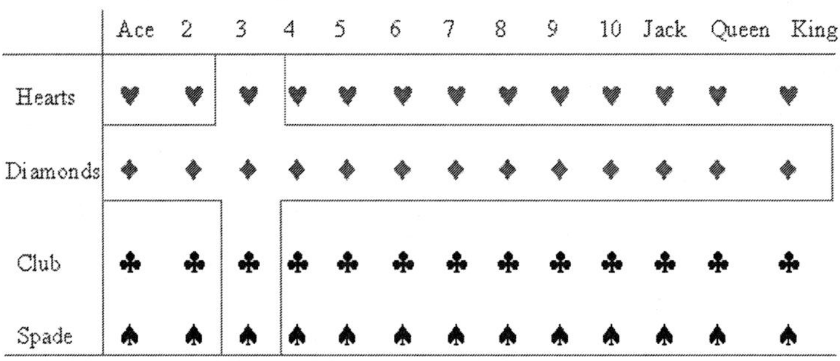

So far we have discussed probabilities of single events. In many situations we come across two or more events occurring together. If event *A* and event *B* are two events and either *A* or *B* or both occurs, is denoted by $A \cap B$ or $(A + B)$ and the event that both *A* and *B* occurs is denoted by $A \cap B$ or *AB*. We term these situations as compound event or the joint occurrence of events. We may need probability that *A* or *B* will happen. It is denoted by $P(A \cap B)$ or $P(A + B)$. Also we may need the probability that *A* and *B* (both) will happen simultaneously. It is denoted by $P(A \cap B)$ or $P(AB)$. Consider a situation, you are asked to choose any 3 or any diamond or both from a well shuffled pack of 52 cards. Now you are interested in the probability of this situation.

It is denoted by $P(A \cup B)$ or $P(A + B)$. Also we may need the probability that A and B (both) will happen simultaneously. It is denoted by $P(A \cap B)$ or $P(AB)$. Consider a situation, you are asked to choose any 3 or any diamond or both from a well shuffled pack of 52 cards. Now you are interested in the probability of this situation.

Now count the dots in the area which fulfills the condition any 3 or any diamond or both. They are 16.

$$= \frac{\text{No. of favorable outcomes}}{\text{Total no. of outcomes}}$$

$$= \frac{16}{52} \text{ or } 0.31$$

Thus the required probability,

In the language of set theory, the set any 3 or any diamond or both is the union of the sets 'any 3 which contains 4 cards ' and 'any diamond' which contains 15 cards. The number of cards in their union is equal to the sum of these numbers minus the number of cards in the space where they overlap. Any points in this space, called the intersection of the two sets, is counted here twice (double counting), once in each set. Dividing by 52 we get the required probability.

$= P(\text{any } 3) + P(\text{any diamond}) - P(3 \text{ of diamond})$

$$= \frac{4}{52} + \frac{13}{52} - \frac{1}{52}$$

$$= \frac{16}{32} \text{ or } 0.31$$

Thus P(any 3 or any diamond or both)

In general, if the letters A and B stands for any two events, then,

$$\left.\begin{cases} P(A \cup B) = P(A) + P(B) - P(A \cap B) \\ \qquad OR \\ P(A + B) + P(A) + P(B) - P(AB) \end{cases}\right\} \text{Addition Law}$$

Clearly, the outcomes of both A and B are non-mutually exclusive.

Example : Two dice are rolled. Find the probability that the score is an even number or multiple of 3.

Solution: Two dice are rolled.

Sample space = {(1, 1), (1, 2), ..., (6, 6)}

$n(S) = 6 \times 6 = 36$

Event E: The score is an even number or multiple of 3.

Note here score means the sum of the numbers on both the dice when they land. For example (1, 1) has score 1 + 1 = 2.It is clear that the least score is 2 and the highest score (6, 6) $\Rightarrow$ 6 + 6 = 12

i.e. score 2, 3, 4, 5, 6, 7, 8, 9, 10, 11, 12

Let Event A: Score is an even numbers,

A = {(1, 1), (1, 3), (1, 5), (2,2), (2, 4), (2, 6), (3, 1), (3, 3) (3, 5), (4, 2), (4, 4), (4, 6), (5, 1), (5, 3), (5, 5), (6, 2), (6, 4), (6, 6)}

Therefore $n(A) = 18$

Let Event B: The score is the multiple of 3

Let Event B: The score is the multiple of 3

i.e. 3, 6, 9, 12.

B = {(1, 2), (1, 5), (2, 4), (2, 1) (3, 6) (3, 3) (4,2), (4, 5), (5, 1), (5,4), (6, 3), (6, 6)}

$n(B) = 12$

Let Event $A \cap B$: The score is an even number and multiple of 3 or (i.e. common to both A and B) AB

AB = {(2, 4), (4, 2), (33,3), (4,2), (5, 1), (6,6)}

$n(AB) = 6$

Now $P(A + B) = P(A) + P(AB)$ where $E = A + B$

$$= \frac{n(A)}{n(S)} + \frac{n(B)}{n(S)} - \frac{n(AB)}{n(S)}$$

$$= \frac{18}{36} + \frac{12}{36} - \frac{6}{36}$$

$$= \frac{14}{36} \quad = \frac{2}{3} \text{ or } 0.67 \text{ (approximately)}$$

Corollary

You are flipping a coin and wanting to find the probability of at least one head or at least one tail. Now it is clear that the probability of one coin flip landing heads is 1/2 and that of one coin flip landing tails is also 1/2. Now these two events are mutually exclusive as you cannot expect a coin land both heads and tails in one trial. Therefore you can determine the required probability by the mere addition of the two probabilities: $\frac{1}{2}+\frac{1}{2}=1$ (or certainty).

Thus if A and B are two mutually exclusive events then the probability that either A or B will happen is the sum of the probabilities of A and B i.e. $P(A + B) = P(A) + P(B)$

Odds In Favour And Against

If P is the probability that an event will occur and $q(= 1 - \text{p})$ is the probability of the non-occurrence of the event; then we say that the odds in favour of the event occurring are $p : q$ (p to q) and the odds against its occurring are $q : p$.

For example, if the event consists of drawing a card if club from the deck of 52 cards, then the odds in favour are, $\frac{13}{52}:\frac{39}{52}$ *i.e.*

$$\frac{1}{4}:\frac{3}{4} \text{ i.e. } 1 : 3;$$

Similarly the odds against a card of diamond would be 3 : 1. The odds in favour of 4 or 6 in a single toss of a fair die are $\frac{1}{3}:\frac{2}{3}$; *i.e.* 1 : 2 and the odds against are 2 : 1.

Multiplication Law of Probability

If there are two independent events; the respective probability of which are known, then the probability that both will happen is the product of the probabilities of their happening respectively $P(AB) = P(A) \times P(B)$

To compute the probability of two or even more independent event all occurring (joint occurrence) extent the above law to required number. For example, first flip a penny, then the nickle and finally flip the dime.

On landing, probability of heads $\frac{1}{2}$ is for a penny probability of heads is $\frac{1}{2}$ for a nickle probability of heads is $\frac{1}{2}$ for a dime,

Thus the probability of landing three heads will be $\frac{1}{2}\times\frac{1}{2}\times\frac{1}{2}=\frac{1}{8}$ or 0.125. (Note that all three events are independent)

Example : Three machines I, II and III manufacture respectively 0.4, 0.5 and 0.1 of the total production. The percentage of defective items produced by I, II and III is 2, 4 and 1 percent respectively for an item randomly chosen, what is the probability it is defective ?

Solution:

The defective items produced by machine,

$$\text{I} = \frac{0.4\times 2}{100}=\frac{8}{100}$$

The defective items produced by machine,

$$\text{II} = \frac{0.5\times 4}{100}=\frac{2}{100}$$

The defective items produced by machine,

$$\text{II} = \frac{0.1\times 1}{100}=\frac{0.1}{100}$$

The total defective items produced by all three machines,

$$= \frac{0.8}{100}+\frac{2}{100}+\frac{0.1}{100}$$

$$= 0.029$$

Thus the requiered probability,

$$= \frac{0.029}{1}$$

$$= 0.029$$

A case of addition Law of mutually exclusive events,

Example : In shuffling a pack of cards, 4 are accidentally dropped one after another. Find the chance that the missing cards should be one from each suit.

Solution: Probability of 4 missing cards from different suits are as follows:

Let H, D, C and S denote heart, diamond, club and spade cards respectively then $P(H) = \frac{13C_1}{52C_1}$

$$P(D) = \frac{13C_1}{52C_1},\ P(C) = \frac{13C_1}{52C_1} \text{ and } P(S) = \frac{13C_1}{52C_1}$$

Since all the four events are independent, using multiplication Law, we get,

Required probability = $P(H) \times P(D) \times P(C) \times P(S)$

$$= \frac{13C_1}{52C_1} \times \frac{13C_1}{52C_1} \times \frac{13C_1}{52C_1} \times \frac{13C_1}{52C_1}$$

$$= \frac{1}{4} \times \frac{1}{4} \times \frac{1}{4} \times \frac{1}{4}$$

$$= \frac{1}{256}$$

Example: A problem in statistics is given to three students A, B and C whose chances in solving it are $\frac{1}{2}, \frac{3}{4}$, and $\frac{1}{4}$ respectively. What is the probability that the problem will be solved?

Solution: The probability that A can solve the problem = 1/2

The probability that B cannot solve the problem = 1 – 1/2 = 1/2

Similarly the probabilities that B and C cannot solve problem are $\left(1-\frac{3}{4}\right)$ and $\left(1-\frac{1}{4}\right)$ respectively.

The probability that A, B, C cannot solve the problem,

$$= \left(1-\frac{1}{2}\right) \left(1-\frac{3}{4}\right) \left(1-\frac{1}{4}\right) \text{.... } (\because \text{ indepemdent events})$$

$$= \frac{1}{2} \times \frac{1}{4} \times \frac{3}{4} = \frac{3}{22}$$

Hence the probability that the problem can be solved,

$$= 1 - \frac{3}{22}$$

$$= \frac{29}{32} = 0.90625$$

Example : Tom and his wife Jenny appear in an interview for two vacancies in the same post. The chance of Tom's selection is 1/7 and that of Jenny's selection is 1/5. What is the probability that (i) both of them will be selected (ii) only one of them will be selected (iii) none of them will be selected.

Solution: P (Tom's selection) = 1/7 P(wife Jenny's selection) = 1/5

1. P(both selected) $= \frac{1}{7} = \frac{1}{5} = \frac{1}{35}$ or 0.03
2. P(only one is selected) = P(only Tom is selected)

   P(Jenny's selection)

$$= \left(\frac{1}{7} \times \frac{4}{5}\right) + \left(\frac{1}{5} \times \frac{6}{7}\right)$$

$$= \frac{10}{35} = \frac{2}{7}$$

Note that when we say only Tom is selected. Simultaneously we mean Jenny is rejected. Thus the two events become independent and hence multiplication rule is applied . Also Tom's selection and Jenny's selection in this case is a mutually exclusive. Hence finally addition rule is applied.

3. P(none of them will be selected) $= \frac{6}{7} \times \frac{4}{5} = \frac{24}{35}$ or 0.69

Conditional Probability

In many situations you get more information than simply the total outcomes and favorable outcomes you already have and, hence you are in position to make yourself more informed to make judgements regarding the probabilities of such situations.

For example, suppose a card is drawn at random from a deck of 52 cards. Let B denotes the event 'the card is a diamond' and A denotes the event 'the card is red'. We may then consider the following probabilities.

1. $P(B) = \frac{13C_1}{52C_1} = \frac{13}{52}$
2. $P(A) = \frac{26C_1}{52C_1} = \frac{26}{52}$

3. $P(AB)$ = The probability that the card is both adiamond and a red card

   $= \frac{13C_1}{52C_1} = \frac{13}{52}$

4. The probability that the card is a diamond knowing that it is red.

Since there are 26 red cards of which 13 are diamonds, the probability that the card is diamond is $\frac{13}{26}$. In other words the probability of event B knowing that A has occurred is $\frac{13}{26}$.

The probability of B under the condition that A has occurred is known as condition probability and it is denoted by $P(B/A)$. Thus $P(B/A) = \frac{13}{26}$. It should be observed that the probability of the event B is increased due to the additional information that the event A has occurred. Conditional probability found using the formula $P(B/A) = \frac{P(AB)}{P(A)}$

Justification:- $P(A/B) = \frac{P(AB)}{P(A)} = \frac{13}{26} = \frac{13/52}{26/52} = \frac{P(AB)}{P(B)}$

Similarly $P(A/B) = \frac{P(AB)}{P(A)}$

In both the cases if A and B are independent events then $P(A/B) = P(A)$ and $P(B/A) = P(B)$

Therefore $P(A) = \frac{P(AB)}{P(B)} \Rightarrow P(AB) = P(A) \, . \, P(B)$

or $P(B) = \frac{P(AB)}{P(A)} \Rightarrow P(AB) = P(A) \, . \, P(B)$

Propositions

- If A and B are independent events then A and B' are also independent where B' is the complementary event B.
- If A and B are independent events then A' and B' are also independent events.

- Two independent events cannot be mutually exclusive.
- De Morgan's Laws: *P(A È B)' = P(A' Ç B')*

  $P(A \cap B)' = P(A' \cup B')$

Example : A bag contains 6 red and 4 blue balls. One ball is drawn at random from the first bag and put into the second bag. A ball is then drawn from the second bag. What is the probability that is red ?

Solution:

- Let A = Getting a red ball from the 1st bag.
- B = Getting a blue from the first bag.
- C = Getting a red ball from the second bag after the ball drawn from the first bag is put into it.

Since there are 10 balls in the first bag out of which 6 are red and 4 are blue.

$$P(A) = \frac{6}{10} \text{ and } P(B) = \frac{4}{10}$$

Suppose that A has occurred. Then a red ball is transferred into second bag. The probability of getting a red ball from the second bag when a red ball is got from the first is $P(C/A) = 5/11$

The probability of getting a red ball from the first bag and a red ball from the second is $P(A \cap C) = P(A) \,.\, P(C/A) =$

$$\frac{6}{10} \times \frac{5}{11} = \frac{30}{110}$$

Now, suppose the event B has occurred then a blue ball is transferred into the second bag. There are now 11 balls in the second bag out of which 4 are red.

$$P(C/B) = \frac{4}{11}$$

$$P(B \cap C) = P(B) \,.\, P(C/B) = \frac{4}{10} \cdot \frac{4}{11} = \frac{16}{110}$$

Since $A \cap C$ and $B \cap C$ are mutually exclusive events

$$P(A \cap C) \cup P(B \cap C) = P(A \cap C) + P(B \cap C)$$

$$= \frac{30}{110} + \frac{16}{110} = \frac{23}{25} \quad i.e.\ 0.92$$

Example : In a certain school, 25% of the boys and 10% of the girls are studying French. The girls constitute 60% of the students boy. If a student is selected at random and is studying French, determine the probability that the student is a girl.

Solution: Let there be 100 students. 60% of which are girls. *i.e.* number of girls is 60 and number of boys is 40.

Then 25% of 40 = $\frac{25}{100} \times 40 = 10$

and,

$$10\% \text{ of } 60 = \frac{10}{100} \times 60 = 6$$

students are studying 'French'

Let A = a student studying French $\Rightarrow n(A) = 16$,

and $P(A) = \frac{n(A)}{n(S)} = \frac{16}{100}$ Let B = a girl student. $A \cap B$ = a girl student studying French,

Now, $n(A \cap B) = 6$, $P(A \cap B) = \frac{n(A \cap B)}{n(S)} = \frac{6}{100}$

Now $P(B/A)$ = Probability that a selected student is a girl studying French

Now $P(B/A) = \frac{P(A \cap B)}{P(A)} = \frac{6/100}{16/100} = \frac{3}{8} = 0.375$

Example : In a school competition, the probability of hitting the target by Dick is 1/2, by Betty is 1/3 and by Joe is 3/5. If all of them fire independently at the same target, calculate the probability that,

- The target will be hit
- Only one of them will hit the target and
- Dick hits, given that only one hits the target.

Solution:

P(Dick hits the target) = 1/2, P(Betty hits the target) = 1/3 and P (Joe hits the target) = 3/5, Now the probabilities of their nothitting the target are

Let P(not-Dick hits the target) = $1 - \frac{1}{2} = \frac{1}{2} = P(D')$

P(not-Betty hits the target) = $1 - \frac{1}{3} = \frac{2}{3} = P(B')$

P(not-Joe hits the target) $= 1-\frac{3}{5}=\frac{2}{5}=P(J')$

- Since original probabilities are independent events,

  $\overline{D}$, $\overline{B}$ and $\overline{J}$ are also independent events.

Hence

$$P(\overline{D}.\overline{B}.\overline{J}) = P(\overline{D}).P(\overline{B}).P(\overline{J})$$

$$= \frac{1}{2}\times\frac{2}{3}\times\frac{2}{5}=\frac{2}{15}$$

The target will be hit if at least one of them hits the target. Therefore, the probability of hitting the target by at least one of them is,

$$1-P(\overline{D}.\overline{B}.\overline{J}) = 1-\frac{2}{15} = \frac{13}{15}$$

- Dick hits but Betty and Joe do not *i.e.* $D\cap\overline{B}\cap\overline{J}$
- Betty hits but not Dick and Joe *i.e.* $\overline{B}\cap D\cap\overline{J}$
- Joe hits but not Dick and Betty *i.e.* $\overline{J}\cap\overline{D}\cap\overline{B}$

Therefore, P(event of hitting the target by only one of Dick, Betty and Joe).

$$P(A)= P(\overline{D})\ P(\overline{B})\ P(\overline{J})+P(\overline{D})\ P(\overline{B}).P(\overline{J})+P(\overline{D})\ P(\overline{B})\ \ P(\overline{J})$$

Chapter 5

Number Theory

Expression

An expression such as "3 + 4 = 5" is nonsense within the scope and definition of real numbers, but it fits nicely within the scope and definition of complex numbers (think of a right triangle with opposite and adjacent sides of 3 and 4, with a hypotenuse of 5). Because complex numbers are two-dimensional, they are able to "add" with one another trigonometrically as single-dimension "real" numbers cannot. Logic is much like mathematics in this respect: the so-called "Laws" of logic depend on how we define what a proposition is. The Greek philosopher Aristotle founded a system of logic based on only two types of propositions: true and false. His bivalent (two-mode) definition of truth led to the four foundational laws of logic: the Law of Identity (A is A); the Law of Non-contradiction (A is not non-A); the Law of the Excluded Middle (either A or non-A); and the Law of Rational Inference.

These so-called Laws function within the scope of logic where a proposition is limited to one of two possible values, but may not apply in cases where propositions can hold values other than "true" or "false." In fact, much work has been done and continues to be done on "multivalued," or *fuzzy* logic, where propositions may be true or false *to a limited degree.*

In such a system of logic, "Laws" such as the Law of the Excluded Middle simply do not apply, because they are founded on the assumption of bivalence. Likewise, many premises which would violate the Law of Non-contradiction in Aristotelian logic have validity in "fuzzy" logic. Again, the defining limits of propositional values determine the Laws describing their functions and relations.

Boole wrote a treatise on the subject in 1854, titled *An Investigation of the Laws of Thought, on Which Are Founded the Mathematical*

Theories of Logic and Probabilities, which codified several rules of relationship between mathematical quantities limited to one of two possible values: true or false, 1 or 0. His mathematical system became known as Boolean algebra.

All arithmetic operations performed with Boolean quantities have but one of two possible outcomes: either 1 or 0. There is no such thing as "2" or "-1" or "1/2" in the Boolean world. It is a world in which all other possibilities are invalid by fiat. As one might guess, this is not the kind of math you want to use when balancing a checkbook or calculating current through a resistor. However, Claude Shannon of MIT fame recognized how Boolean algebra could be applied to on-and-off circuits, where all signals are characterized as either "high" (1) or "low" (0). The thesis, titled *A Symbolic Analysis of Relay and Switching Circuits*, put Boole's theoretical work to use in a way Boole never could have imagined, giving us a powerful mathematical tool for designing and analyzing digital circuits.

In this chapter, you will find a lot of similarities between Boolean algebra and "normal" algebra, the kind of algebra involving so-called real numbers. Just bear in mind that the system of numbers defining Boolean algebra is severely limited in terms of scope, and that there can only be one of two possible values for any Boolean variable: 1 or 0.

Consequently, the "Laws" of Boolean algebra often differ from the "Laws" of real-number algebra, making possible such statements as 1 + 1 = 1, which would normally be considered absurd. Once you comprehend the premise of all quantities in Boolean algebra being limited to the two possibilities of 1 and 0, and the general philosophical principle of Laws depending on quantitative definitions, the "nonsense" of Boolean algebra disappears.It should be clearly understood that Boolean numbers are not the same as *binary* numbers.

Whereas Boolean numbers represent an entirely different system of mathematics from real numbers, binary is nothing more than an alternative *notation* for real numbers. The two are often confused because both Boolean math and binary notation use the same two ciphers: 1 and 0.The difference is that Boolean quantities are restricted to a single bit (either 1 or 0), whereas binary numbers may be composed of many bits adding up in place-weighted form to a value of any finite size. The binary number 10011_2("nineteen") has no more place in the Boolean world than the decimal number 2_{10} ("two") or the octal number 32_8 ("twenty-six").

Boolean Arithmetic

In computer science, the Boolean or logical data type is a primitive data type having one of two values: true or false, intended to represent the truth values of logic and Boolean algebra. In programming languages that have a built-in Boolean data type, such as Pascal and Java, the comparison operators such as'>' and'?' are usually defined to return a Boolean value. Also, conditional and iterative commands may be defined to test Boolean-valued expressions.

Languages without an explicit Boolean data type, like C and Lisp, may still represent truth values it by some other data type. Lisp uses an empty list for false, and any other value for true. C uses an integer type, with false represented as the zero value, and true as any non-zero value (such as 1 or -1). Indeed, a Boolean variable may be regarded (and be implemented) as a numerical variable with a single binary digit (bit), which can store only two values. Let us begin our exploration of Boolean algebra by adding numbers together:

$$0 + 0 = 0$$

$$0 + 1 = 1$$

$$1 + 0 = 1$$

$$1 + 1 = 1$$

The first three sums make perfect sense to anyone familiar with elementary addition. The last sum, though, is quite possibly responsible for more confusion than any other single statement in digital electronics, because it seems to run contrary to the basic principles of mathematics. Well, it *does* contradict principles of addition for real numbers, but not of elimination. It does not matter how many or few terms we add together, either.

Consider the following sums:

$0 + 1 + 1 = 1$

$1 + 1 + 1 = 1$

$0 + 1 + 1 + 1 = 1$

$1 + 0 + 1 + 1 + 1 = 1$

Take a close look at the two-term sums in the first set of equations. Does that pattern look familiar to you? It should! It is the same pattern of 1's and 0's as seen in the truth table for an OR gate. In other words, Boolean addition corresponds to the logical function of an "OR" gate, as well as to parallel switch contacts:

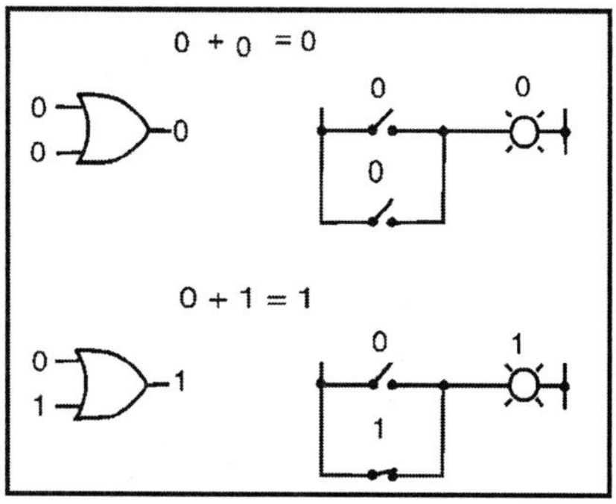

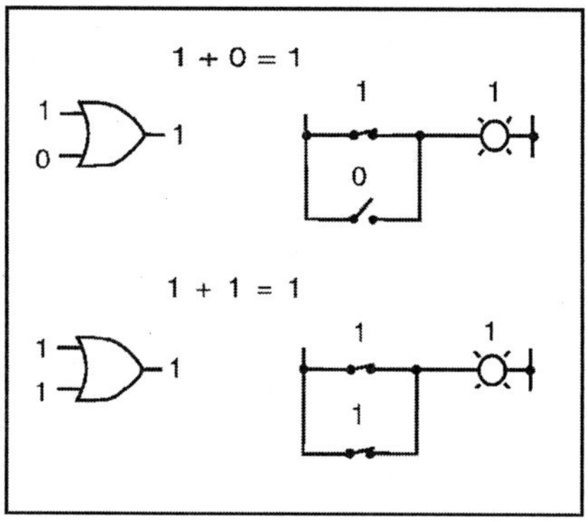

There is no such thing as subtraction in the realm of Boolean mathematics. Subtraction implies the existence of negative numbers: 5–3 is the same thing as 5 + (–3), and in Boolean algebra negative quantities are forbidden. There is no such thing as division in Boolean mathematics, either, since division is really nothing more than compounded subtraction, in the same way that multiplication is compounded addition.

Multiplication is valid in Boolean algebra, and thankfully it is the same as in real-number algebra: anything multiplied by 0 is 0, and anything multiplied by 1 remains unchanged:

$$0 \times 0 = 0$$

$$0 \times 1 = 0$$

$$1 \times 0 = 0$$

$$1 \times 1 = 1$$

This set of equations should also look familiar to you: it is the same pattern found in the truth table for an AND gate. In other words, Boolean multiplication corresponds to the logical function of an "AND" gate, as well as to series switch contacts:

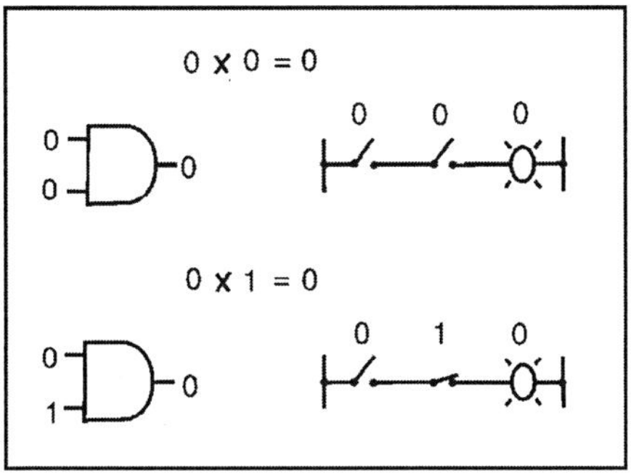

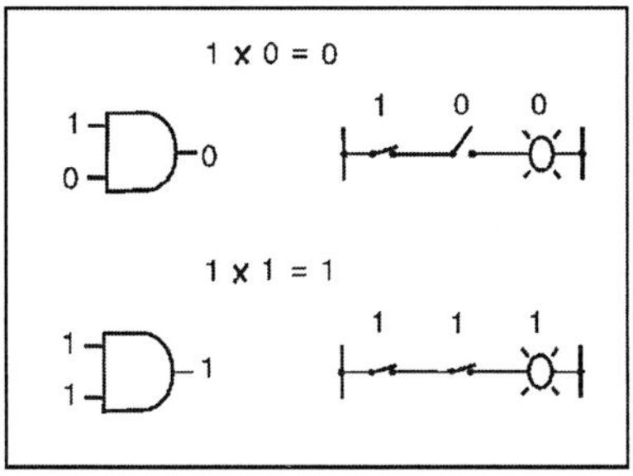

Like "normal" algebra, Boolean algebra uses alphabetical letters to denote variables. Unlike "normal" algebra, though, Boolean variables are always CAPITAL letters, never lower-case. Because they are allowed to possess only one of two possible values, either 1 or 0, each and every variable has a*complement*: the opposite of its value.

For example, if variable "A" has a value of 0, then the complement of A has a value of 1. Boolean notation uses a bar above the variable character to denote complementation, like this:

If: A = 0

Then: $\overline{A}=1$

If: A = 1

Then: $\overline{A}=1$

In written form, the complement of "A" denoted as "A-not" or "A-bar". Sometimes a "prime" symbol is used to represent complementation. For example, A' would be the complement of A, much the same as using a prime symbol to denote differentiation in calculus rather than the fractional notation d/dt. Usually, though, the "bar" symbol finds more widespread use than the "prime" symbol, for reasons that will become more apparent later in this chapter.

Boolean complementation finds equivalency in the form of the NOT gate, or a normally-closed switch or relay contact:

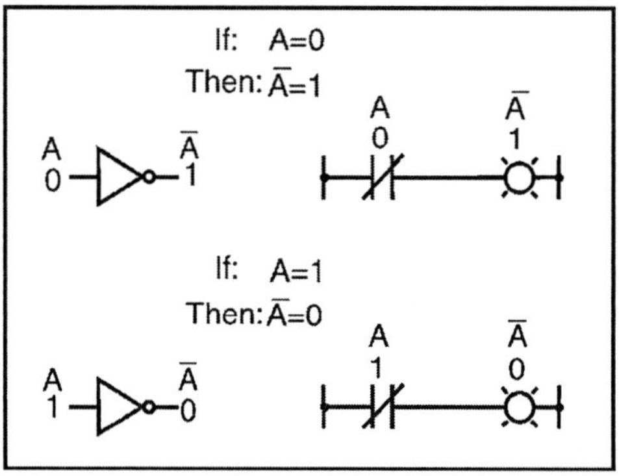

The basic definition of Boolean quantities has led to the simple rules of addition and multiplication, and has excluded both subtraction and division as valid arithmetic operations. We have a symbology for denoting Boolean variables, and their complements. In the next section we will proceed to develop Boolean identities.

- *Review:*
- Boolean addition is equivalent to the *OR* logic function, as well as *parallel* switch contacts.
- Boolean multiplication is equivalent to the *AND* logic function, as well as *series* switch contacts.
- Boolean complementation is equivalent to the *NOT* logic function, as well as *normally-closed*relay contacts.

Boolean Algebraic Identities

In mathematics, an *identity* is a statement true for all possible values of its variable or variables. The algebraic identity of x + 0 = x tells us that anything (x) added to zero equals the original "anything," no matter what value that "anything" (x) may be. Like ordinary algebra, Boolean algebra has its own unique identities based on the bivalent states of Boolean variables. The first Boolean identity is that the sum of anything and zero is the same as the original "anything." This identity is no different from its real-number algebraic equivalent:

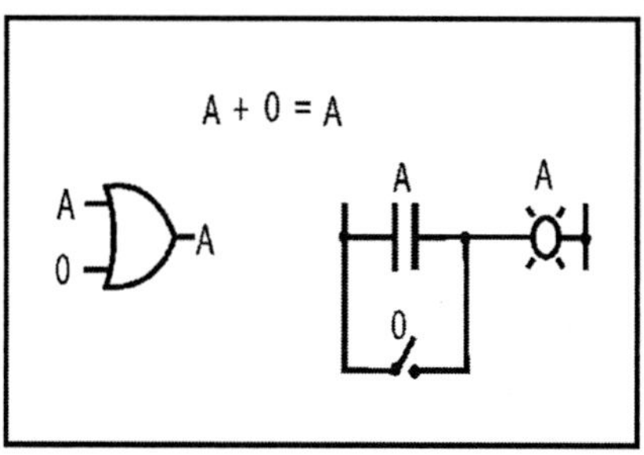

No matter what the value of A, the output will always be the same: when A=1, the output will also be 1; when A=0, the output will also be 0. The next identity is most definitely *different* from any seen in normal algebra. Here we discover that the sum of anything and one is one:

$$A + 1 = 1$$

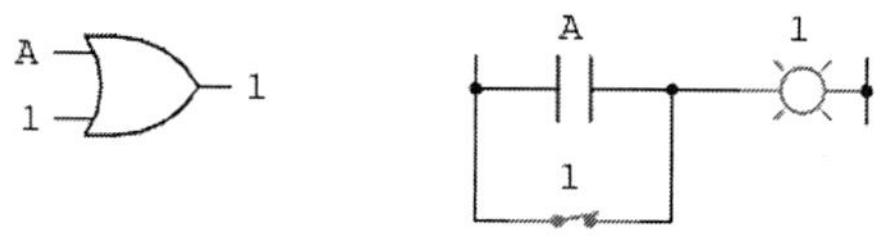

No matter what the value of A, the sum of A and 1 will always be 1. In a sense, the "1" signal *overrides*the effect of A on the logic circuit, leaving the output fixed at a logic level of 1.Next, we examine the effect of adding A and A together, which is the same as connecting both inputs of an OR gate to each other and activating them with the same signal:

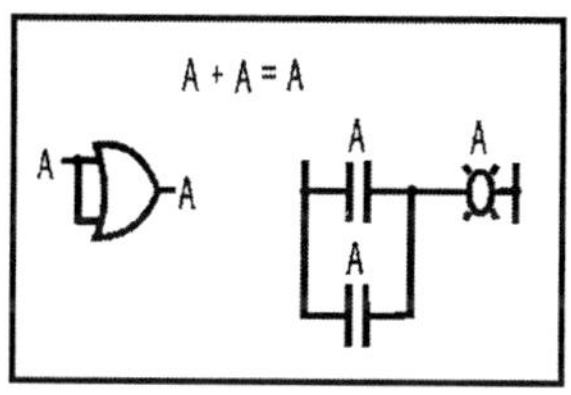

In real-number algebra, the sum of two identical variables is twice the original variable's value (x + x = 2x), but remember that there is no concept of "2" in the world of Boolean math, only 1 and 0, so we cannot say that A + A = 2A. Thus, when we add a Boolean quantity to itself, the sum is equal to the original quantity:

$$0 + 0 = 0,$$

and

$$1 + 1 = 1.$$

Introducing the uniquely Boolean concept of complementation into an additive identity, we find an interesting effect. Since there must be one "1" value between any variable and its complement, and since the sum of any Boolean quantity and 1 is 1, the sum of a variable and its complement must be 1:

$$A + \bar{A} = 1$$

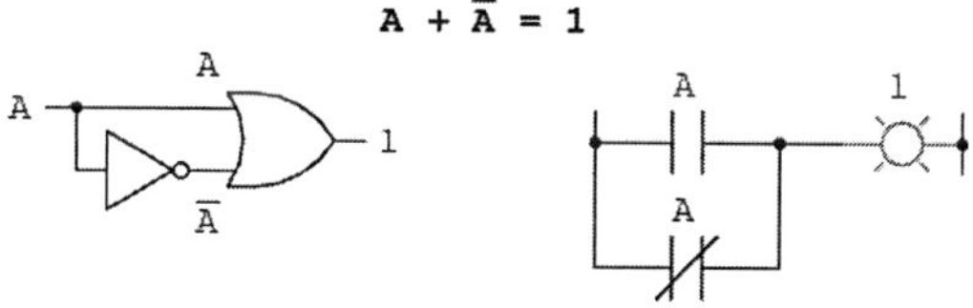

Just as there are four Boolean additive identities (A+0, A+1, A+A, and A+A'), so there are also four multiplicative identities: A × 0, A × 1, A × A, and A × A'. Of these, the first two are no different from their equivalent expressions in regular algebra:

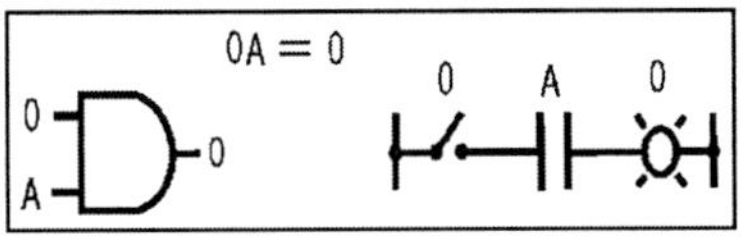

1A = A

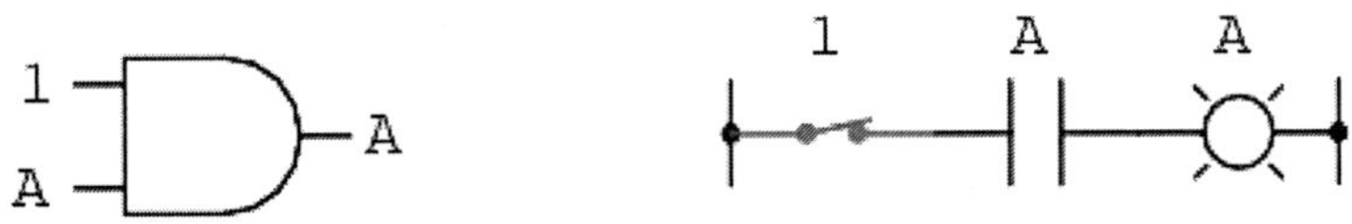

The third multiplicative identity expresses the result of a Boolean quantity multiplied by itself. In normal algebra, the product of a variable and itself is the *square* of that variable ($3 \times 3 = 3^2 = 9$). However, the concept of "square" implies a quantity of 2, which has no meaning in Boolean algebra, so we cannot say that $A \times A = A^2$. Instead, we find that the product of a Boolean quantity and itself is the original quantity, since $0 \times 0 = 0$ and $1 \times 1 = 1$:

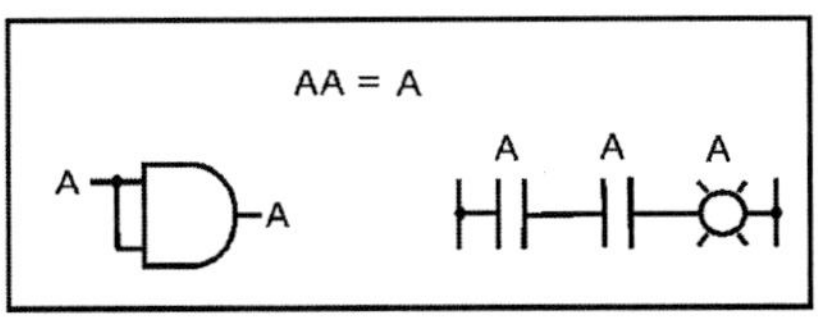

The fourth multiplicative identity has no equivalent in regular algebra because it uses the complement of a variable, a concept unique to Boolean mathematics. Since there must be one "0" value between any variable and its complement, and since the product of any Boolean quantity and 0 is 0, the product of a variable and its complement must be 0:

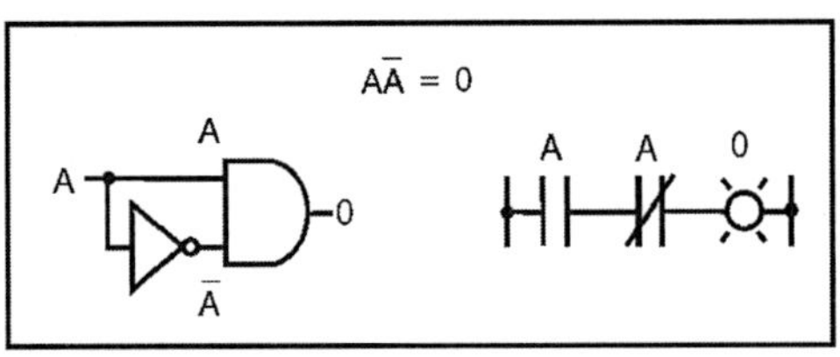

To summarize, then, we have four basic Boolean identities for addition and four for multiplication:

Table: *Basic Boolean Algebraic Identities*

Additive	*Multiplicative*
$A + 0 = A$	$0A = 0$
$A + 1 = 1$	$1A = A$
$A + A = A$	$AA = A$
$A + \overline{A} = 1$	$A\overline{A} = 0$

Another identity having to do with complementation is that of the *double complement*: a variable inverted twice. Complementing a variable twice (or any even number of times) results in the original Boolean value. This is analogous to negating (multiplying by -1) in real-number algebra: an even number of negations cancel to leave the original value:

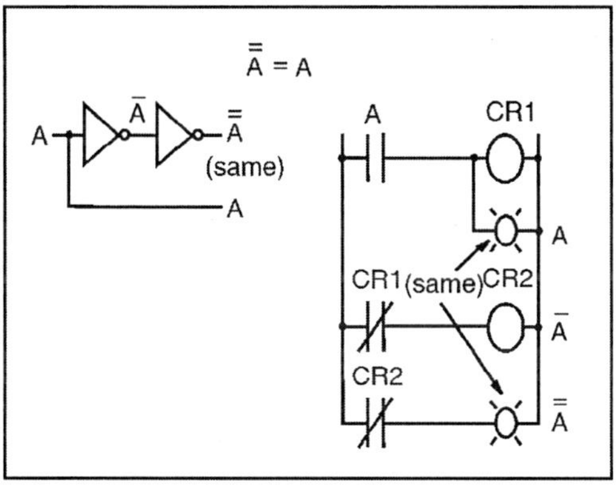

Boolean Algebraic Properties

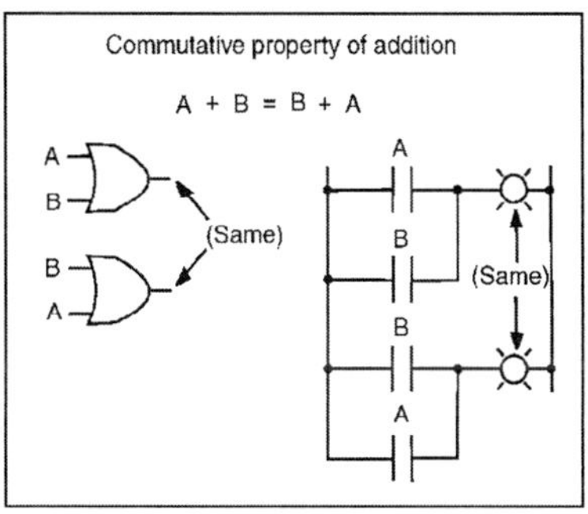

Another type of mathematical identity, called a "property" or a "law," describes how differing variables relate to each other in a system of numbers. One of these properties is known as the *commutative property*, and it applies equally to addition and multiplication. In essence, the commutative property tells us we can reverse the order of variables that are either added together or multiplied together without changing the truth of the expression:

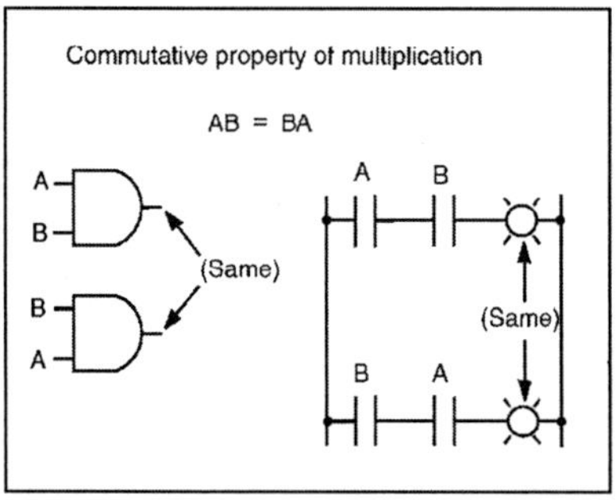

Along with the commutative properties of addition and multiplication, we have the *associative property*, again applying equally well to addition and multiplication. This property tells us we can associate groups of added or multiplied variables together with parentheses without altering the truth of the equations.

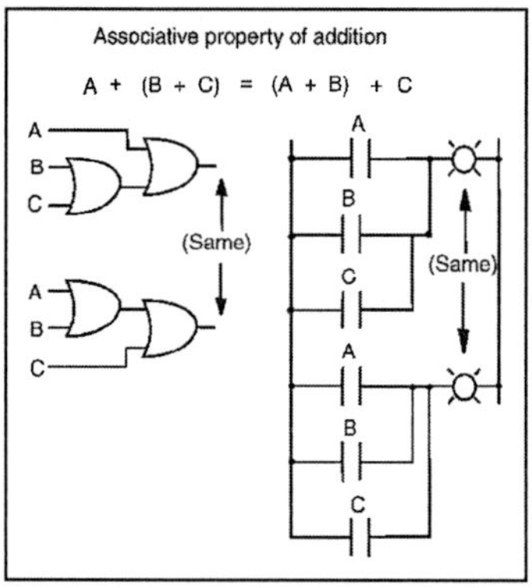

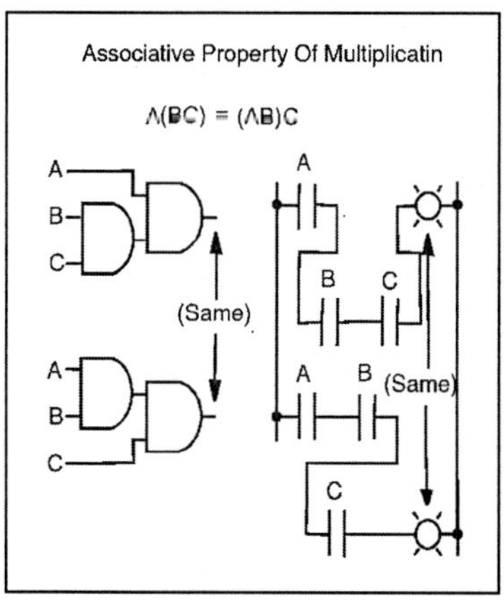

Lastly, we have the *distributive property*, illustrating how to expand a Boolean expression formed by the product of a sum, and in reverse shows us how terms may be factored out of Boolean sums-of-products:

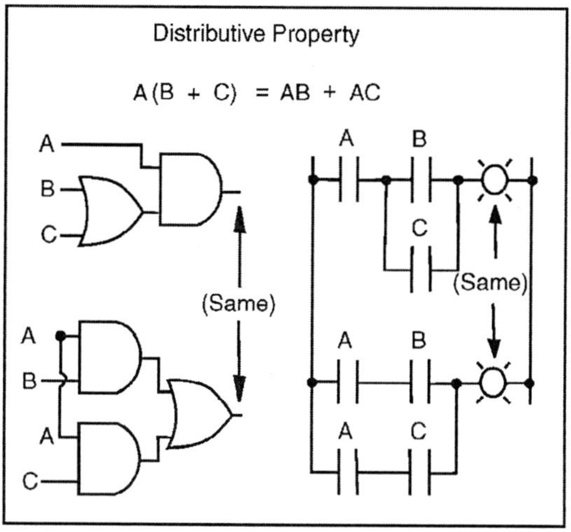

To summarize, here are the three basic properties: commutative, associative, and distributive.

Table: Basic Bookean Algebraic Properties

Additive	*Multiplicative*
A + B = B + A	AB = BA
A + (B + C) = (A + B) + C	A(BC) = (AB) C
A (B + C) = AB + AC	

Boolean Simplification Rules

Boolean algebra finds its most practical use in the simplification of logic circuits. If we translate a logic circuit's function into symbolic (Boolean) form, and apply certain algebraic rules to the resulting equation to reduce the number of terms and/or arithmetic operations, the simplified equation may be translated back into circuit form for a logic circuit performing the same function with fewer components.

If equivalent function may be achieved with fewer components, the result will be increased reliability and decreased cost of manufacture.To this end, there are several rules of Boolean algebra presented in this section for use in reducing expressions to their simplest forms.

The identities and properties already reviewed in this chapter are very useful in Boolean simplification, and for the most part bear similarity to many identities and properties of "normal" algebra. However, the rules shown in this section are all unique to Boolean mathematics.

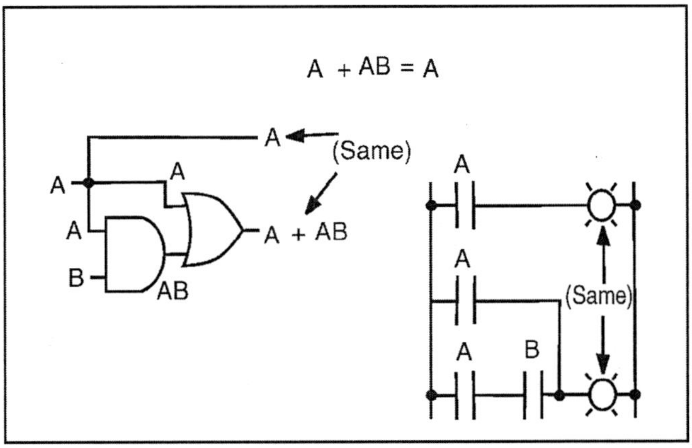

This rule may be proven symbolically by factoring an "A" out of the two terms, then applying the rules of A + 1 = 1 and 1A = A to achieve the final result:

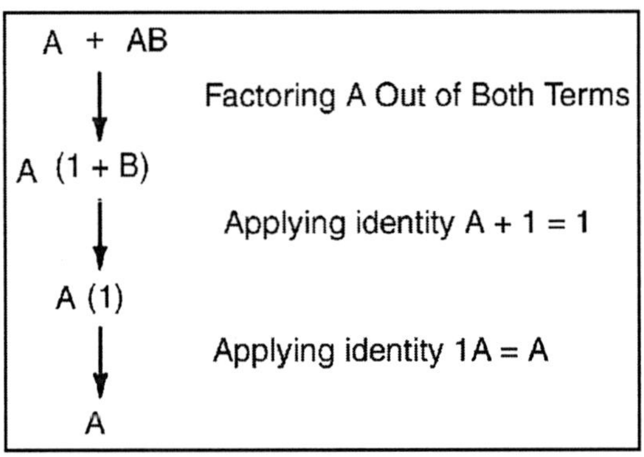

Please note how the rule A + 1 = 1 was used to reduce the (B + 1) term to 1. When a rule like "A + 1 = 1" is expressed using the letter "A", it doesn't mean it only applies to expressions containing "A". What the "A" stands for in a rule like A + 1 = 1 is *any* Boolean variable or collection of variables. This is perhaps the most difficult concept for new students to master in Boolean simplification: applying standardized identities, properties, and rules to expressions not in standard form. For instance, the Boolean expression ABC + 1 also reduces to 1 by means of the "A + 1 = 1" identity. In this case, we recognize that the "A" term in the identity's standard form can represent the entire "ABC" term in the original expression.

The next rule looks similar to the first one shown in this section, but is actually quite different and requires a more clever proof:

$$A + \bar{A}B = A + B$$

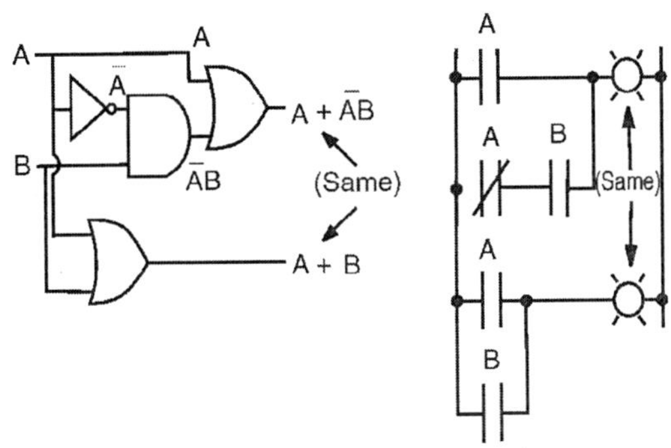

$A + \bar{A}B$

↓ Applying the Previous Rule to Expand A Term $A + AB = A$

$A + AB + \bar{A}B$

↓ Factoring B Out of 2nd and 3rd Terms

$A + B(A + \bar{A})$

↓ Applying Identity A + A = 1

$A + B(1)$

↓ Applying Identity 1A = A

$A + B$

Note how the last rule (A + AB = A) is used to "un-simplify" the first "A" term in the expression, changing the "A" into an "A + AB". While this may seem like a backward step, it certainly helped to reduce the expression to something simpler!

Sometimes in mathematics we must take "backward" steps to achieve the most elegant solution. Knowing when to take such a step and when not to is part of the art-form of algebra, just as a victory in a game of chess almost always requires calculated sacrifices.

Another rule involves the simplification of a product-of-sums expression:

(A + B) (A + C) = A + BC

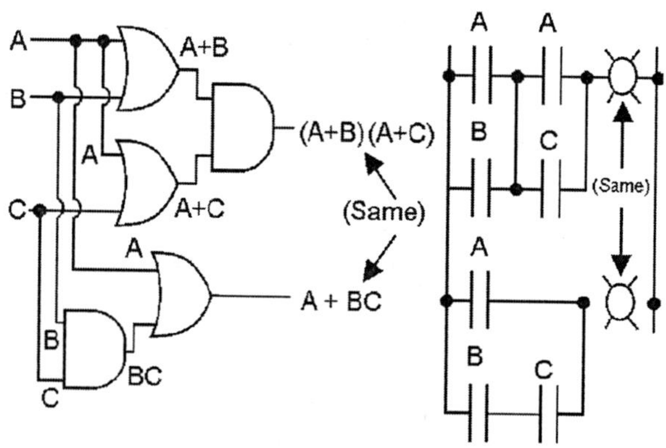

And, the corresponding proof:

(B + C) (A + C)

↓ Distributing Terms

AA + AC + AB + BC

↓ Applying Identity AA = A

A + AC + AB + BC

↓ Applying Rule A + AB = A to the A + AC term

A + AB + BC

↓ Applying Rule A + AB = A to the A + AB term

A + BC

To summarize, here are the three new rules of Boolean simplification expounded in this section:

Useful Boolean Rules for Simplification:

$$A + AB = A$$

$$A + \overline{A}B = A + B$$

$$(A + B)(A + C) = A + BC$$

Circuit Simplification

Let's begin with a semiconductor gate circuit in need of simplification. The "A," "B," and "C" input signals are assumed to be provided from switches, sensors, or perhaps other gate circuits. Where these signals originate is of no concern in the task of gate reduction.

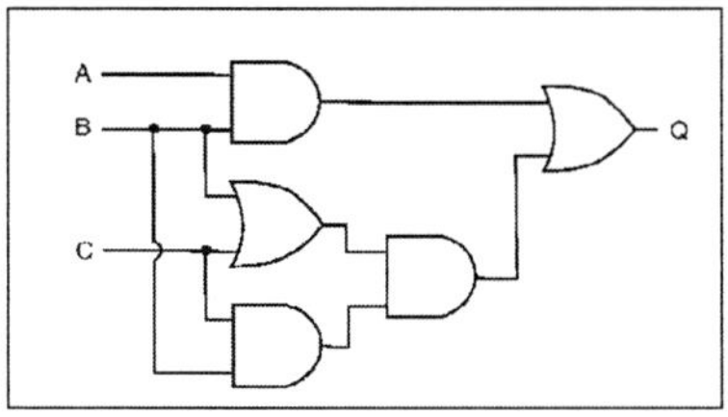

Our first step in simplification must be to write a Boolean expression for this circuit. This task is easily performed step by step if we start by writing sub-expressions at the output of each gate, corresponding to the respective input signals for each gate. Remember that OR gates are equivalent to Boolean addition, while AND gates are equivalent to Boolean multiplication. For example, I'll write sub-expressions at the outputs of the first three gates:

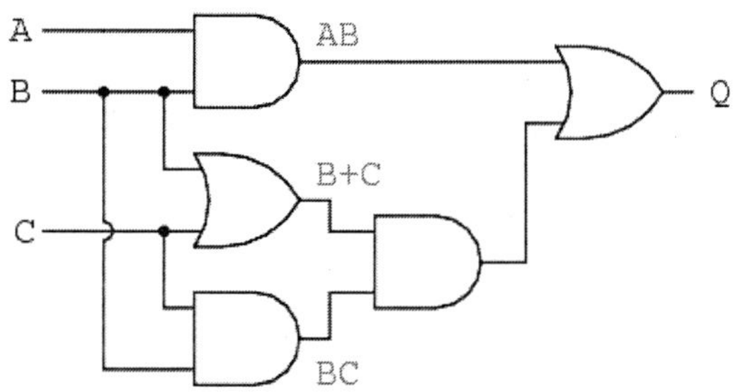

then another sub-expression for the next gate:

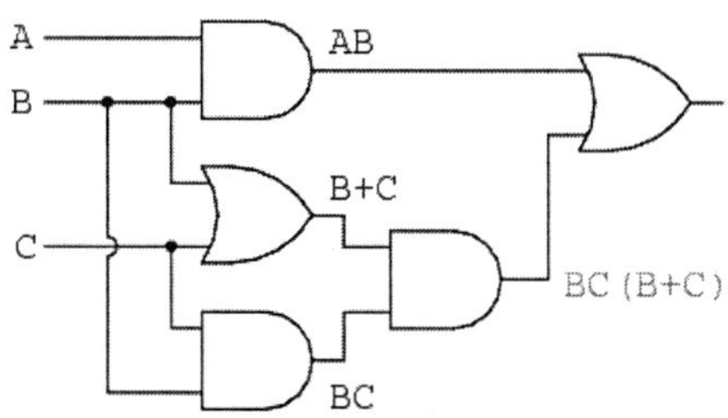

Finally, the output ("Q") is seen to be equal to the expression AB + BC(B + C):

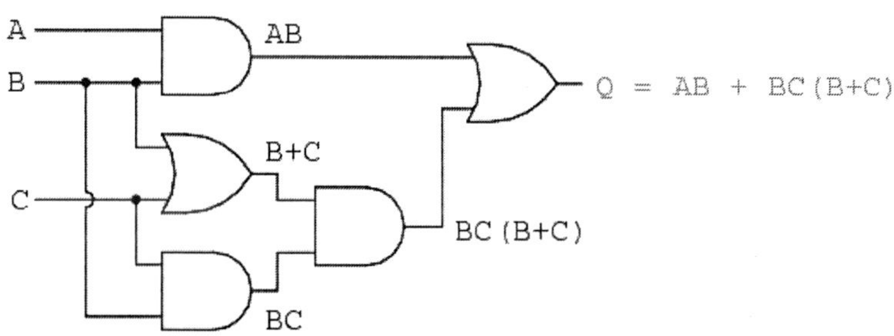

Now that we have a Boolean expression to work with, we need to apply the rules of Boolean algebra to reduce the expression to its simplest form (simplest defined as requiring the fewest gates to implement):

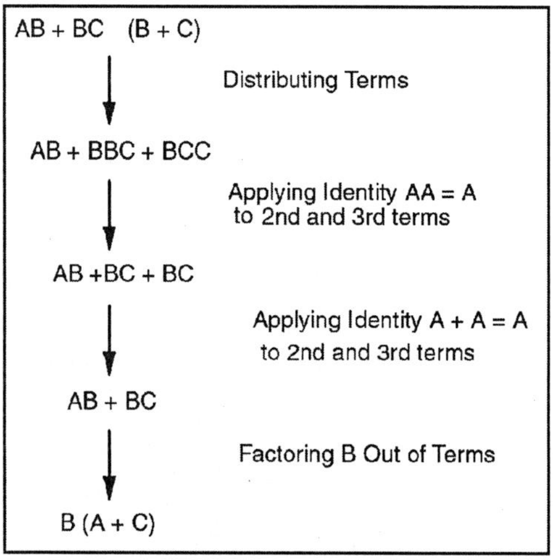

The final expression, B(A + C), is much simpler than the original, yet performs the same function. If you would like to verify this, you may generate a truth table for both expressions and determine Q's status (the circuits' output) for all eight logic-state combinations of A, B, and C, for both circuits. The two truth tables should be identical. Now, we must generate a schematic diagram from this Boolean expression. To do this, evaluate the expression, following proper mathematical order of operations (multiplication before addition, operations inside parentheses before anything else), and draw gates for each step.

Remember again that OR gates are equivalent to Boolean addition, while AND gates are equivalent to Boolean multiplication. In this case, we would begin with the sub-expression "A + C", which is an OR gate:

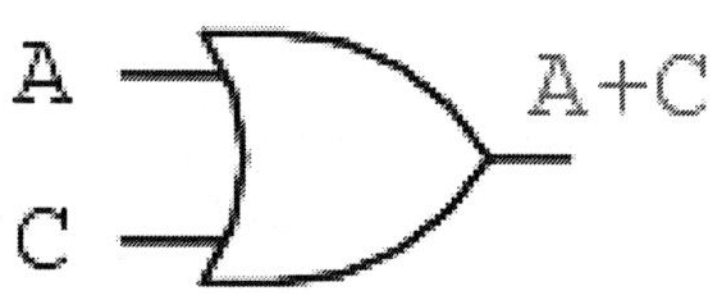

The next step in evaluating the expression "B(A + C)" is to multiply (AND gate) the signal B by the output of the previous gate (A + C):

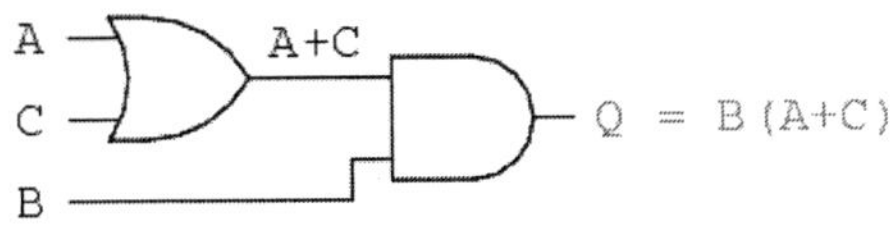

Obviously, this circuit is much simpler than the original, having only two logic gates instead of five. Such component reduction results in higher operating speed (less delay time from input signal transition to output signal transition), less power consumption, less cost, and greater reliability.

Electromechanical relay circuits, typically being slower, consuming more electrical power to operate, costing more, and having a shorter average life than their semiconductor counterparts, benefit dramatically from Boolean simplification. Let's consider an example circuit:

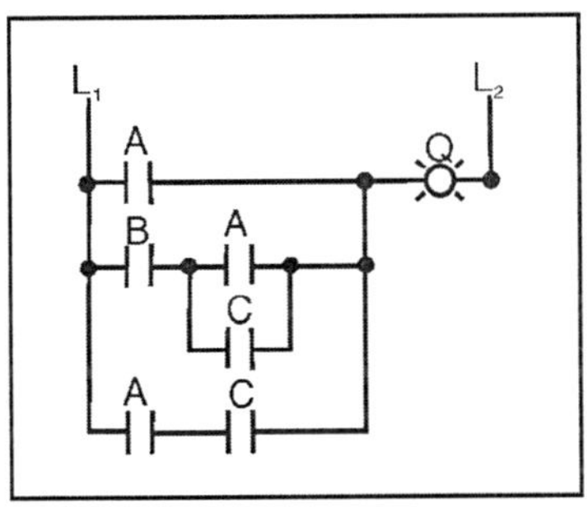

As before, our first step in reducing this circuit to its simplest form must be to develop a Boolean expression from the schematic. The easiest way I've found to do this is to follow the same steps I'd normally follow to reduce a series-parallel resistor network to a single, total resistance.

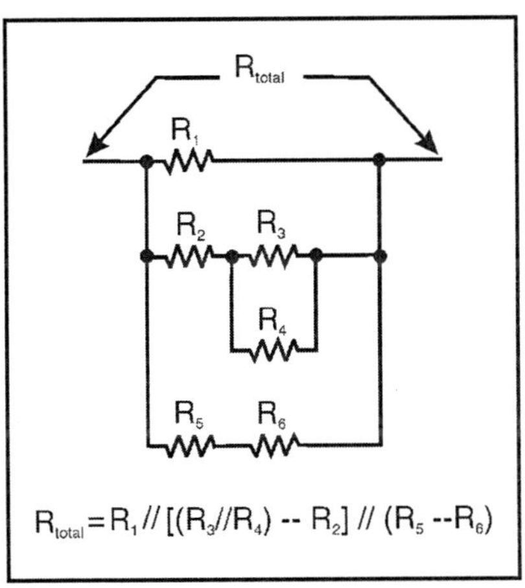

For example, examine the following resistor network with its resistors arranged in the same connection pattern as the relay contacts in the former circuit, and corresponding total resistance formula:

Remember that parallel contacts are equivalent to Boolean addition, while series contacts are equivalent to Boolean multiplication. Write a Boolean expression for this relay contact circuit, following the same order of precedence that you would follow in reducing a series-parallel resistor network to a total resistance. It may be helpful to write a Boolean sub-expression to the left of each ladder “rung,” to help organize your expression-writing:

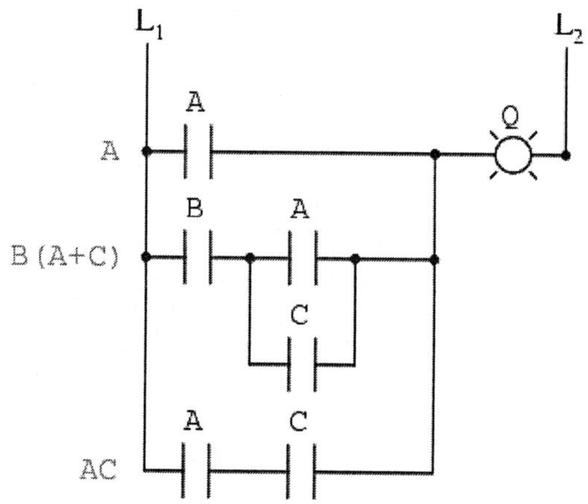

$$Q = A + B(A+C) + AC$$

Now that we have a Boolean expression to work with, we need to apply the rules of Boolean algebra to reduce the expression to its simplest form (simplest defined as requiring the fewest relay contacts to implement):

$A + B(A + C) + AC$

↓ Distributing Terms

$A + AB + BC + AC$

↓ Applying Rule $A + AB = A$ to 1st and 2nd terms

$A + BC + AC$

↓ Applying Rule $A + AB = A$ to 1st and 3rd terms

$A + BC$

The more mathematically inclined should be able to see that the two steps employing the rule "A + AB = A" may be combined into a single step, the rule being expandable to:

"A + AB + AC + AD +... = A"

A + B (A + C) + AC

↓ Distributing Terms

A + AB + BC + AC

↓ Applying (Expanded) Rule A + AB = A to 1st, 2nd, and 4th terms

A + BC

As you can see, the reduced circuit is much simpler than the original, yet performs the same logical function:

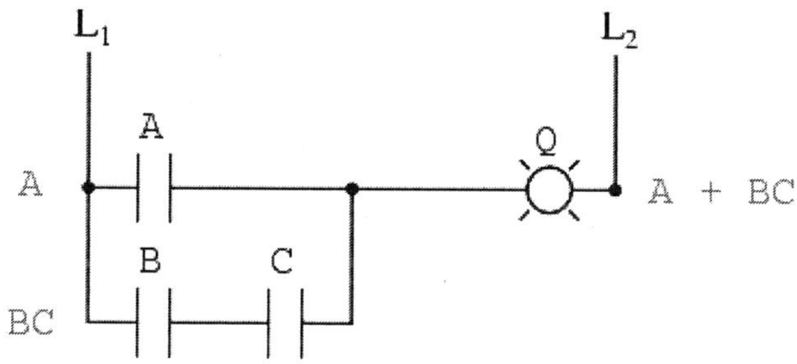

Review: To convert a gate circuit to a Boolean expression, label each gate output with a Boolean sub-expression corresponding to the gates' input signals, until a final expression is reached at the last gate.

To convert a Boolean expression to a gate circuit, evaluate the expression using standard order of operations: multiplication before addition, and operations within parentheses before anything else.

To convert a ladder logic circuit to a Boolean expression, label each rung with a Boolean sub-expression corresponding to the contacts' input signals, until a final expression is reached at the last coil or light. To determine proper order of evaluation, treat the contacts as though they were resistors, and as if you were determining total resistance of the series-parallel network formed by them.

In other words, look for contacts that are either *directly* in series or *directly* in parallel with each other first, then "collapse" them into equivalent Boolean sub-expressions before proceeding to other contacts.

To convert a Boolean expression to a ladder logic circuit, evaluate the expression using standard order of operations: multiplication before addition, and operations within parentheses before anything else.

Logical Boolean Algebra

George Boole believed in what he called the 'process of analysis', that is, the process by which combinations of interpretable symbols are obtained. It is the use of these symbols according to well-determined methods of combination that he believed presented 'true calculus'.

Today, all our computers use Boole's logic system - using microchips that contain thous ands of tiny electronic switches arranged into logical 'gates' that produce predictable and reliable conclusions.

The basic logic gates are AND, OR and NOT. It is these gates, used in different combinations, that allow the computer to execute its operations using binary language.Each gate assesses information (consisting of high or low voltages) in accordance with predetermined rules, and produces a single high or low voltage logical conclusion.

The voltage itself represents the binary yes-no, true-false, one-zero concept.

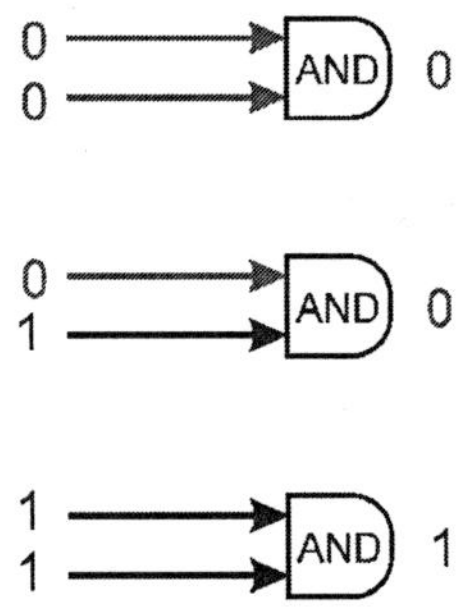

AND gates will only yield a TRUE result (that is, a binary 1) if all input is TRUE. Therefore, the top two gates will produce a FALSE (binary 0) result.

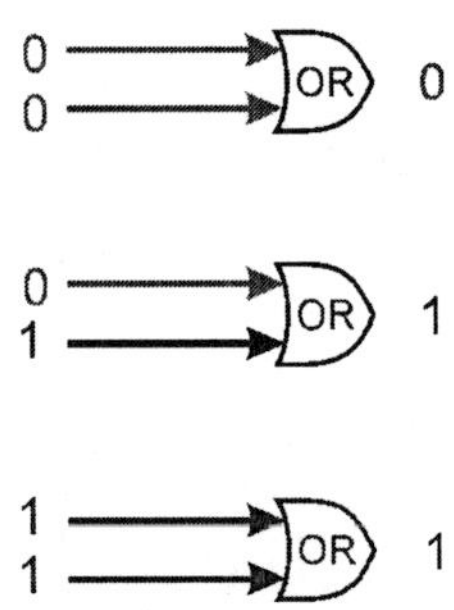

OR gates are less fussy. An OR gate will send a TRUE result if any of its input is TRUE. Only if every input is FALSE will it produce a 0 result.

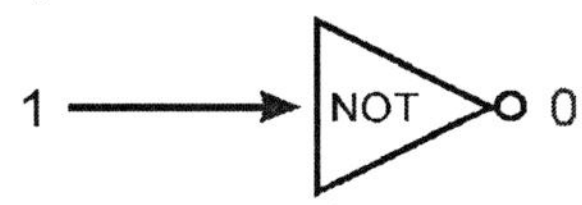

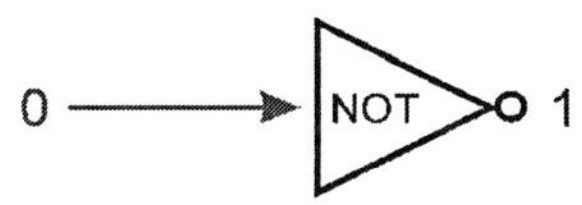

NOT gates, on the other hand, do just one thing. A NOTgate accepts only a single input, either TRUE or FALSE, which it promptly reverses. NOT gates can be combined with AND or OR to give NAND (not and) orNOR (not or) gates. These new gates process input in the usual manner and then reverse the result.

Putting them all together produces something like this ('this', by the way, is known as a half-adder). The top gate inputs 1 and 0 through OR to give a 1 result. The bottom gate channels 1 and 0 through AND to produce 0, which is inverted through the NOTgate into a second 1. The last AND gate sees two inputs of 1 and promptly spits out a TRUE (1) result. Thus, 1 + 0 = 1.

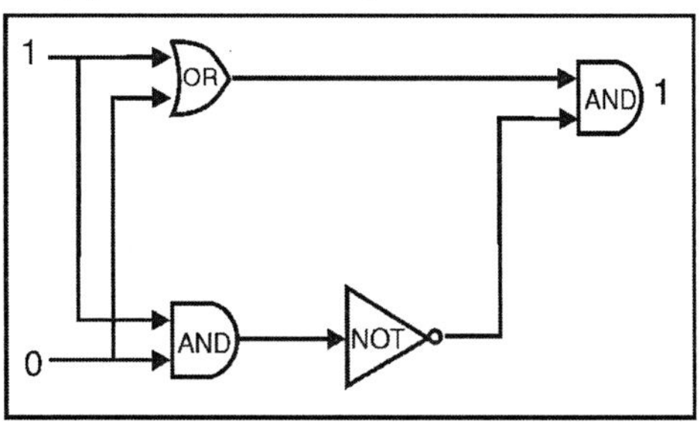

By adjusting the gates, binary adders can be used not only for addition, but for subtraction, multiplication, and division as well, enabling them to control the computer's data flow and to process instructions.

The Exclusive-OR function

One element conspicuously missing from the set of Boolean operations is that of Exclusive-OR. Whereas the OR function is equivalent to Boolean addition, the AND function to Boolean multiplication, and the NOT function (inverter) to Boolean complementation, there is no direct Boolean equivalent for Exclusive-OR. This hasn't stopped people from developing a symbol to represent it, though:

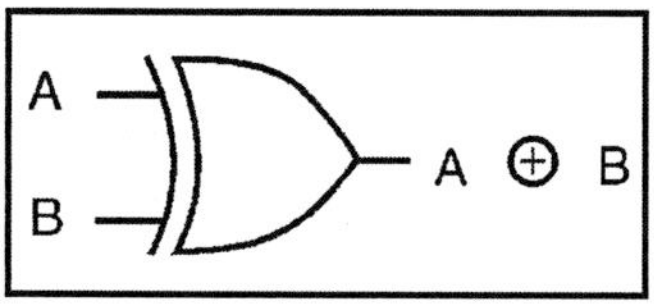

This symbol is seldom used in Boolean expressions because the identities, laws, and rules of simplification involving addition, multiplication, and complementation do not apply to it. However, there is a way to represent the Exclusive-OR function in terms of OR and AND, as has been shown in previous chapters: AB' + A'B

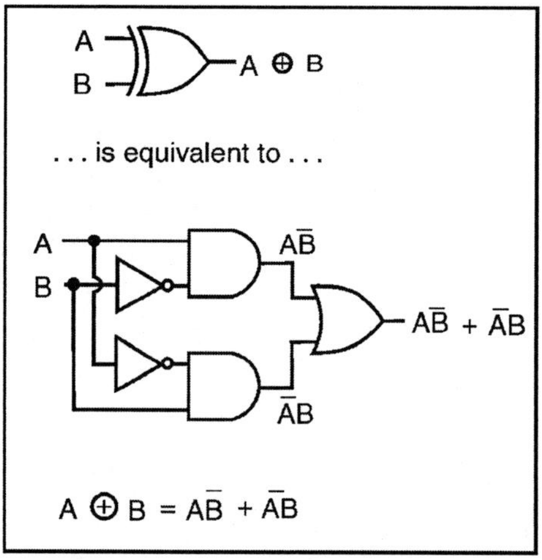

As a Boolean equivalency, this rule may be helpful in simplifying some Boolean expressions. Any expression following the AB' + A'B form (two AND gates and an OR gate) may be replaced by a single Exclusive-OR gate.

DeMorgan's Theorems

A mathematician named DeMorgan developed a pair of important rules regarding group complementation in Boolean algebra. By *group* complementation, I'm referring to the complement of a group of terms, represented by a long bar over more than one variable.

You should recall from the chapter on logic gates that inverting all inputs to a gate reverses that gate's essential function from AND to OR, or vice versa, and also inverts the output. So, an OR gate with all inputs inverted (a Negative-OR gate) behaves the same as a

NAND gate, and an AND gate with all inputs inverted (a Negative-AND gate) behaves the same as a NOR gate. DeMorgan's theorems state the same equivalence in "backward" form: that inverting the output of any gate results in the same function as the opposite type of gate (AND vs. OR) with inverted inputs:

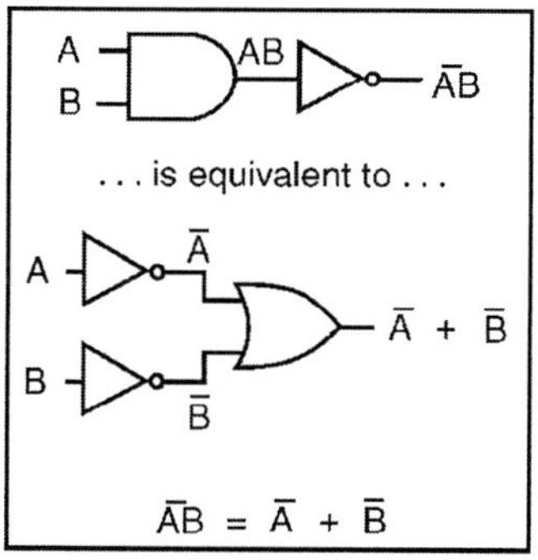

A long bar extending over the term AB acts as a grouping symbol, and as such is entirely different from the product of A and B independently inverted. In other words, (AB)' is not equal to A'B'. Because the "prime" symbol (') cannot be stretched over two variables like a bar can, we are forced to use parentheses to make it apply to the whole term AB in the previous sentence. A bar, however, acts as its own grouping symbol when stretched over more than one variable. This has profound impact on how Boolean expressions are evaluated and reduced, as we shall see.

DeMorgan's theorem may be thought of in terms of *breaking* a long bar symbol. When a long bar is broken, the operation directly underneath the break changes from addition to multiplication, or vice versa, and the broken bar pieces remain over the individual variables. To illustrate:

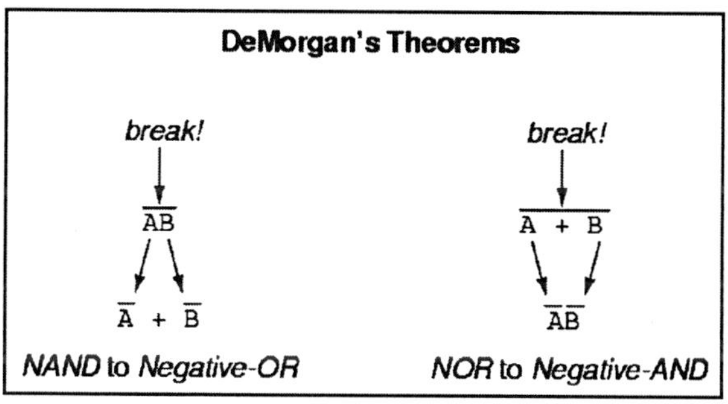

When multiple "layers" of bars exist in an expression, you may only break *one bar at a time*, and it is generally easier to begin simplification by breaking the longest (uppermost) bar first. To illustrate, let's take the expression (A + (BC)')' and reduce it using DeMorgan's Theorems:

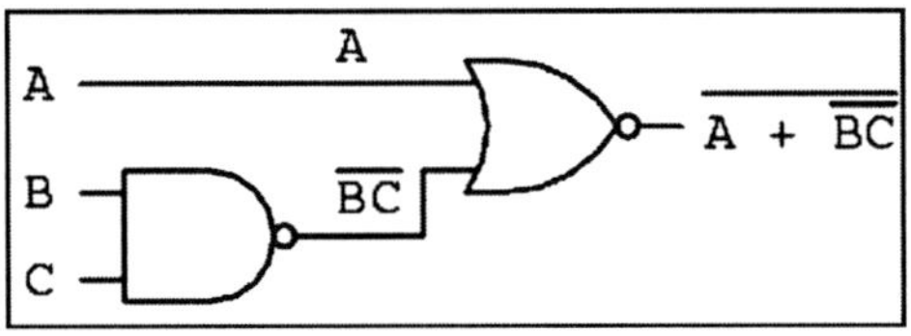

Following the advice of breaking the longest (uppermost) bar first, I'll begin by breaking the bar covering the entire expression as a first step:

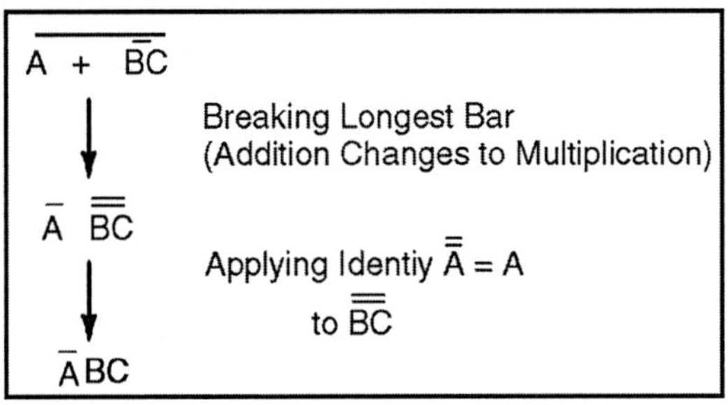

As a result, the original circuit is reduced to a three-input AND gate with the A input inverted:

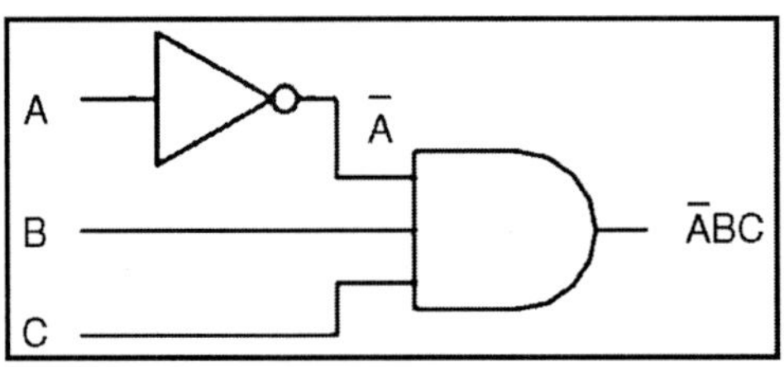

You should *never* break more than one bar in a single step, as illustrated here:

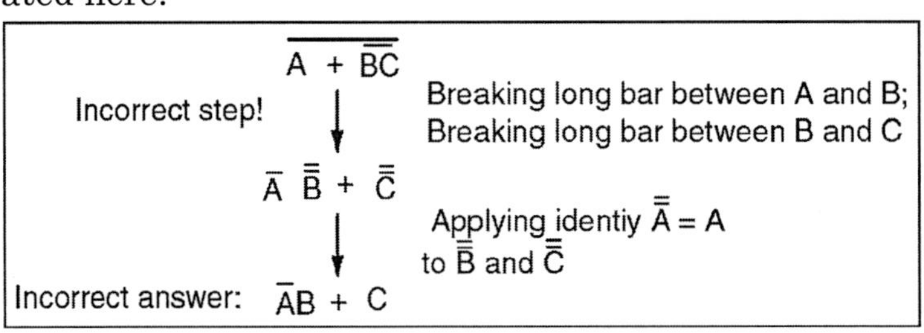

As tempting as it may be to conserve steps and break more than one bar at a time, it often leads to an incorrect result, so don't do it!

It is possible to properly reduce this expression by breaking the short bar first, rather than the long bar first:

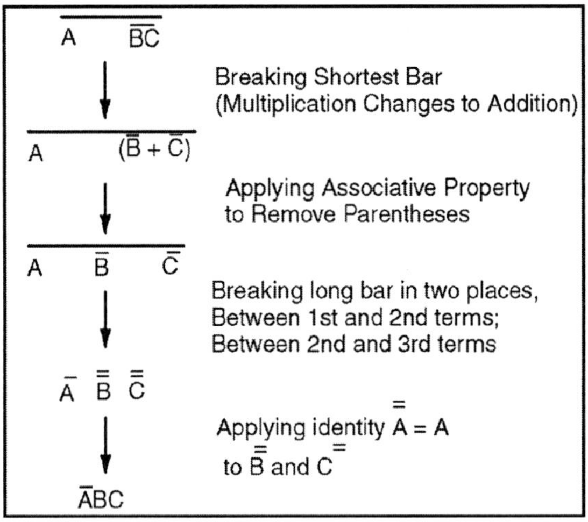

The end result is the same, but more steps are required compared to using the first method, where the longest bar was broken first. Note how in the third step we broke the long bar in two places. This is a legitimate mathematical operation, and not the same as breaking two bars in one step!

The prohibition against breaking more than one bar in one step is *not* a prohibition against breaking a bar in more than one place. Breaking in more than one *place* in a single step is okay; breaking more than one *bar* in a single step is not.You might be wondering why parentheses were placed around the sub-expression B' + C', considering the fact that I just removed them in the next step.

I did this to emphasize an important but easily neglected aspect of DeMorgan's theorem. Since a long bar functions as a grouping symbol, the variables formerly grouped by a broken bar must remain grouped lest proper precedence (order of operation) be lost. In this example, it really wouldn't matter if I forgot to put parentheses in after breaking the short bar, but in other cases it might. Consider this example, starting with a different expression:

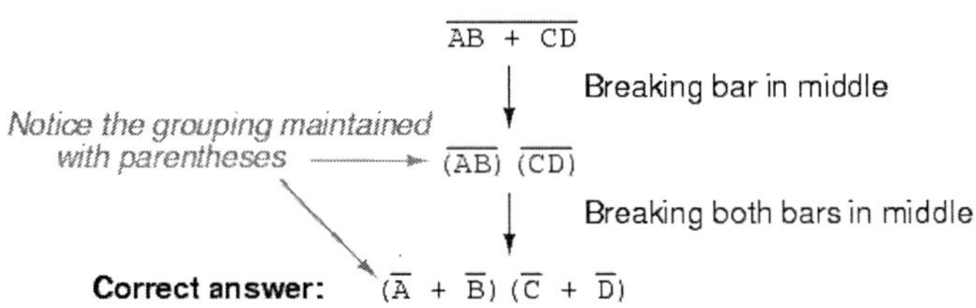

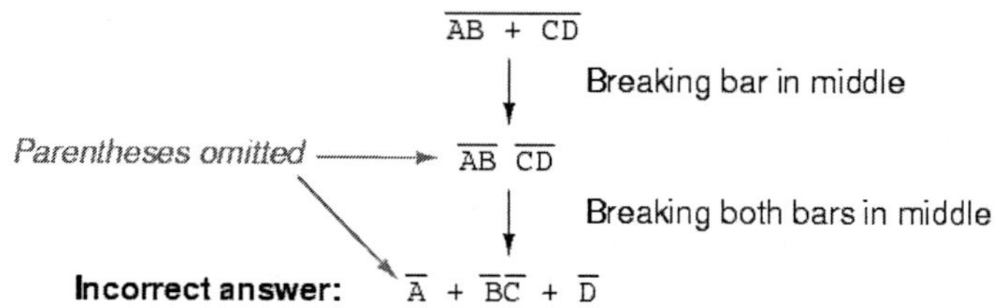

As you can see, maintaining the grouping implied by the complementation bars for this expression is crucial to obtaining the correct answer.Let's apply the principles of DeMorgan's theorems to the simplification of a gate circuit:

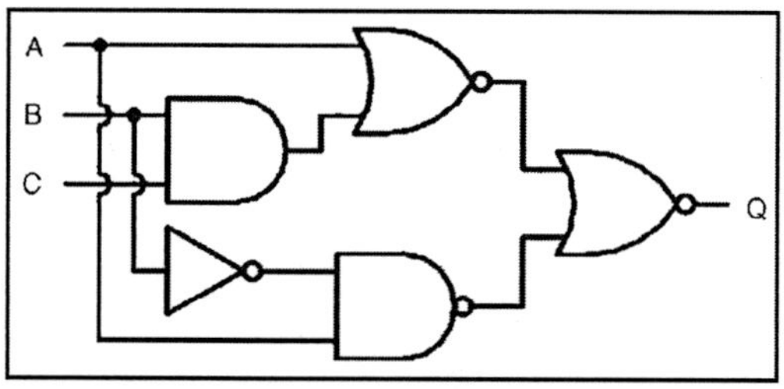

As always, our first step in simplifying this circuit must be to generate an equivalent Boolean expression. We can do this by placing a sub-expression label at the output of each gate, as the inputs become known. Here's the first step in this process:

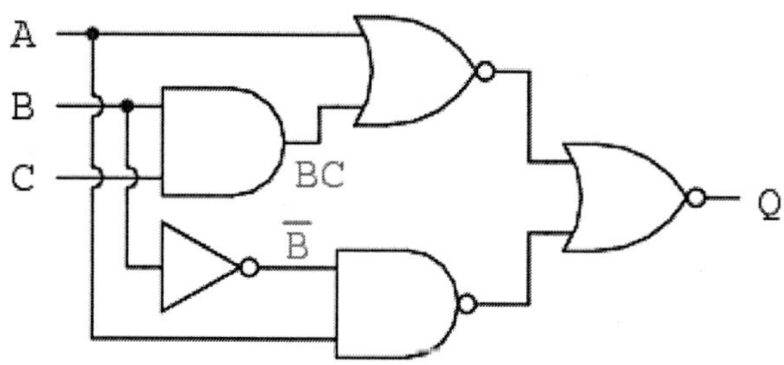

Next, we can label the outputs of the first NOR gate and the NAND gate. When dealing with inverted-output gates, I find it easier to write an expression for the gate's output *without* the final inversion, with an arrow pointing to just before the inversion bubble.

Then, at the wire leading out of the gate (after the bubble), I write the full, complemented expression.

This helps ensure I don't forget a complementing bar in the sub-expression, by forcing myself to split the expression-writing task into two steps:

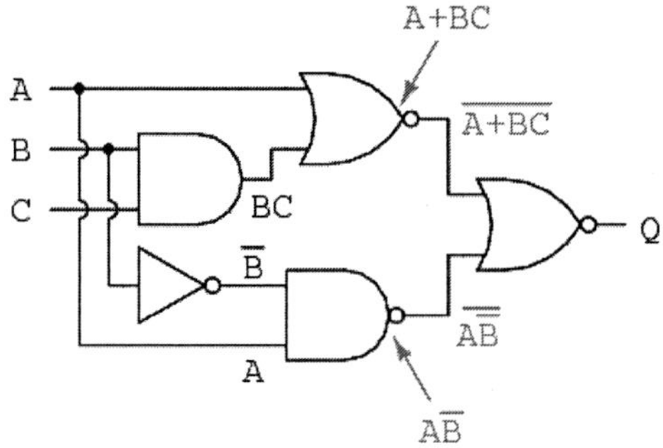

Finally, we write an expression (or pair of expressions) for the last NOR gate:

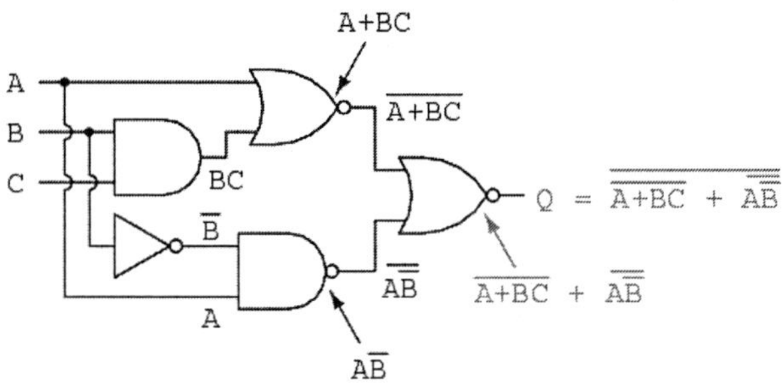

Now, we reduce this expression using the identities, properties, rules, and theorems (DeMorgan's) of Boolean algebra:

$$\overline{\overline{A+BC} + \overline{A\overline{B}}}$$

↓ Breaking Longest Bar

$$\overline{\overline{(A+BC)}}\quad \overline{\overline{(A\overline{B})}}$$

↓ Applying Identity $\overline{\overline{A}} = A$ Wherever Double Bars of Equal Length are Found

$$(A+BC)\quad (A\overline{B})$$

↓ Distributive Property

$$AA\overline{B} + BCA\overline{B}$$

↓ Applying Identity AA = A to left Term; Applying Identity $A\overline{A} = 0$ to B and $\overline{B}$ in Right Term

$$A\overline{B} + 0$$

↓ Applying Identity A + 0 = A

$$A\overline{B}$$

The equivalent gate circuit for this much-simplified expression is as follows:

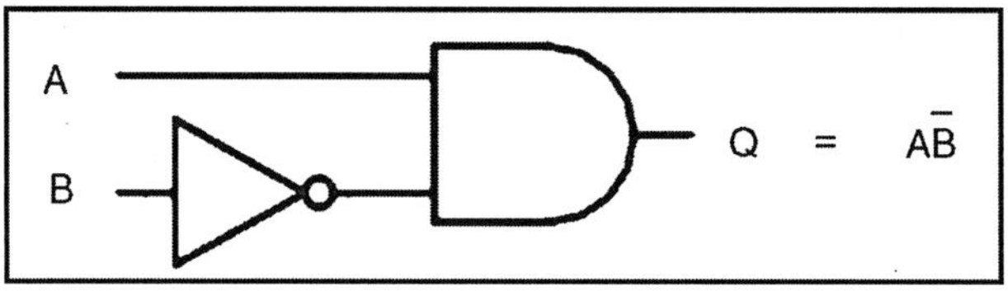

Review

DeMorgan's Theorems describe the equivalence between gates with inverted inputs and gates with inverted outputs. Simply put, a NAND gate is equivalent to a Negative-OR gate, and a NOR gate is equivalent to a Negative-AND gate.

When "breaking" a complementation bar in a Boolean expression, the operation directly underneath the break (addition or multiplication) reverses, and the broken bar pieces remain over the respective terms. It is often easier to approach a problem by breaking the longest (uppermost) bar before breaking any bars under it. You must *never* attempt to break two bars in one step!

Complementation bars function as grouping symbols. Therefore, when a bar is broken, the terms underneath it must remain grouped. Parentheses may be placed around these grouped terms as a help to avoid changing precedence.

Converting Truth tables into Boolean expressions

In designing digital circuits, the designer often begins with a truth table describing what the circuit should do. The design task is largely to determine what type of circuit will perform the function described in the truth table.

While some people seem to have a natural ability to look at a truth table and immediately envision the necessary logic gate or relay logic circuitry for the task, there are procedural techniques available for the rest of us. Here, Boolean algebra proves its utility in a most dramatic way.To illustrate this procedural method, we should begin with a realistic design problem.

Suppose we were given the task of designing a flame detection circuit for a toxic waste incinerator. The intense heat of the fire is intended to neutralize the toxicity of the waste introduced into the incinerator. Such combustion-based techniques are commonly used to neutralize medical waste, which may be infected with deadly viruses or bacteria:

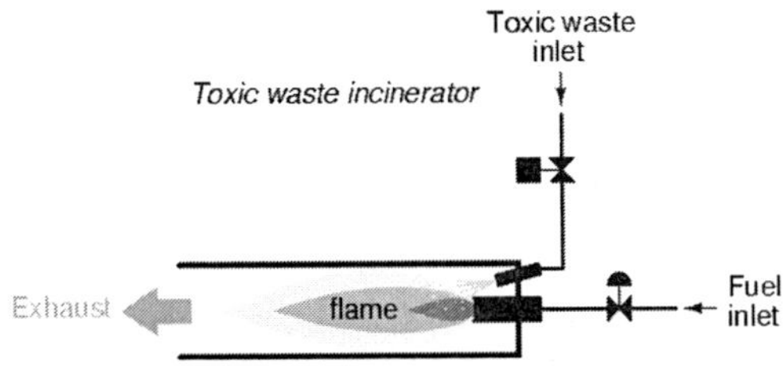

So long as a flame is maintained in the incinerator, it is safe to inject waste into it to be neutralized. If the flame were to be extinguished, however, it would be unsafe to continue to inject waste into the combustion chamber, as it would exit the exhaust un-neutralized, and pose a health threat to anyone in close proximity to the exhaust.What we need in this system is a sure way of detecting the presence of a flame, and permitting waste to be injected only if a flame is "proven" by the flame detection system. Several different flame-detection technologies exist: optical (detection of light), thermal (detection of high temperature), and electrical conduction (detection of ionized particles in the flame path), each one with its unique advantages and disadvantages.

Suppose that due to the high degree of hazard involved with potentially passing un-neutralized waste out the exhaust of this incinerator, it is decided that the flame detection system be made redundant (multiple sensors), so that failure of a single sensor does not lead to an emission of toxins out the exhaust.ach sensor comes equipped with a normally-open contact (open if no flame, closed if flame detected) which we will use to activate the inputs of a logic system:

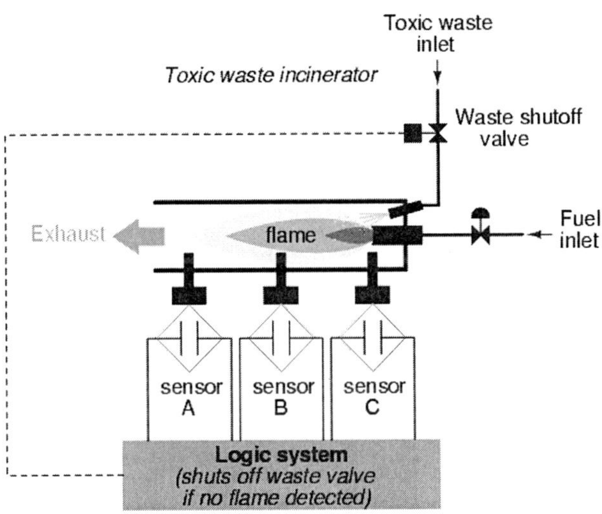

Our task, now, is to design the circuitry of the logic system to open the waste valve if and only if there is good flame proven by the sensors. First, though, we must decide what the logical behaviour of this control system should be. Do we want the valve to be opened if only one out of the three sensors detects flame? Probably not, because this would defeat the purpose of having multiple sensors.

If any one of the sensors were to fail in such a way as to falsely indicate the presence of flame when there was none, a logic system based on the principle of "any one out of three sensors showing flame" would give the same output that a single-sensor system would with the same failure.

A far better solution would be to design the system so that the valve is commanded to open if any only if *all three sensors*detect a good flame. This way, any single, failed sensor falsely showing flame could not keep the valve in the open position; rather, it would require all three sensors to be failed in the same manner— a highly improbable scenario—for this dangerous condition to occur.

Thus, our truth table would look like this:

sensor inputs

A	B	C	Output
0	0	0	0
0	0	1	0
0	1	0	0
0	1	1	0
1	0	0	0
1	0	1	0
1	1	0	0
1	1	1	1

Output = 0 (close valve)

Output = 1 (open valve)

It does not require much insight to realise that this functionality could be generated with a three-input AND gate: the output of the circuit will be "high" if and only if input A *AND* input B *AND* input C are all "high:"

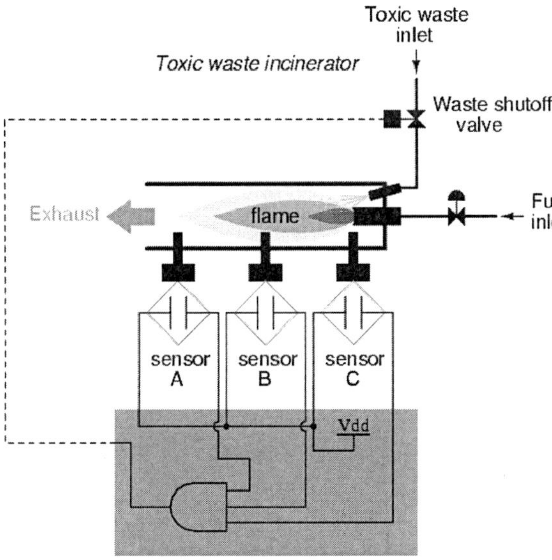

If using relay circuitry, we could create this AND function by wiring three relay contacts in series, or simply by wiring the three sensor contacts in series, so that the only way electrical power could be sent to open the waste valve is if all three sensors indicate flame:

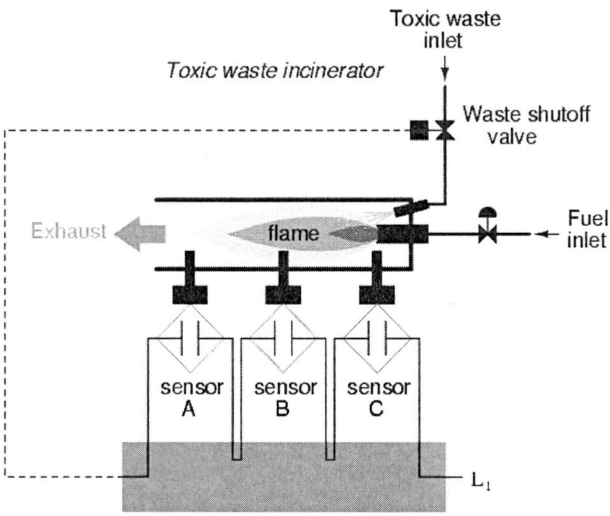

While this design strategy maximizes safety, it makes the system very susceptible to sensor failures of the opposite kind. Suppose that one of the three sensors were to fail in such a way that it indicated no flame when there really was a good flame in the incinerator's combustion chamber. That single failure would shut off the waste valve unnecessarily, resulting in lost production time and wasted fuel (feeding a fire that wasn't being used to incinerate waste).

It would be nice to have a logic system that allowed for this kind of failure without shutting the system down unnecessarily, yet still provide sensor redundancy so as to maintain safety in the event that any single sensor failed "high" (showing flame at all times, whether or not there was one to detect). A strategy that would meet both needs would be a "two out of three" sensor logic, whereby the waste valve is opened if at least two out of the three sensors show good flame. The truth table for such a system would look like this:

sensor inputs

A	B	C	Output
0	0	0	0
0	0	1	0
0	1	0	0
0	1	1	1
1	0	0	0
1	0	1	1
1	1	0	1
1	1	1	1

Output = 0 (close valve)

Output = 1 (open valve)

Here, it is not necessarily obvious what kind of logic circuit would satisfy the truth table. However, a simple method for designing such a circuit is found in a standard form of Boolean expression called the *Sum-Of-Products*, or *SOP*, form. As you might suspect, a Sum-Of-Products Boolean expression is literally a set of Boolean terms added (*summed*) together, each term being a multiplicative (*product*) combination of Boolean variables. An example of an SOP expression would be something like this: ABC + BC + DF, the sum of products "ABC," "BC," and "DF."

Sum-Of-Products expressions are easy to generate from truth tables. All we have to do is examine the truth table for any rows where the output is "high" (1), and write a Boolean product term that would equal a value of 1 given those input conditions. For instance, in the fourth row down in the truth table for our two-out-of-three logic system, where A=0, B=1, and C=1, the product term would be A'BC, since that term would have a value of 1 if and only if A=0, B=1, and C=1:

sensor inputs

A	B	C	Output	
0	0	0	0	
0	0	1	0	
0	1	0	0	
0	1	1	1	$\overline{A}BC = 1$
1	0	0	0	
1	0	1	1	
1	1	0	1	
1	1	1	1	

Three other rows of the truth table have an output value of 1, so those rows also need Boolean product expressions to represent them:

sensor inputs

A	B	C	Output	
0	0	0	0	
0	0	1	0	
0	1	0	0	
0	1	1	1	$\overline{A}BC = 1$
1	0	0	0	
1	0	1	1	$A\overline{B}C = 1$
1	1	0	1	$AB\overline{C} = 1$
1	1	1	1	$ABC = 1$

Finally, we join these four Boolean product expressions together by addition, to create a single Boolean expression describing the truth table as a whole:

sensor inputs

A	B	C	Output	
0	0	0	0	
0	0	1	0	
0	1	0	0	
0	1	1	1	$\overline{A}BC = 1$
1	0	0	0	
1	0	1	1	$A\overline{B}C = 1$
1	1	0	1	$AB\overline{C} = 1$
1	1	1	1	$ABC = 1$

$$\text{Output} = \overline{A}BC + A\overline{B}C + AB\overline{C} + ABC$$

Now that we have a Boolean Sum-Of-Products expression for the truth table's function, we can easily design a logic gate or relay logic circuit based on that expression:

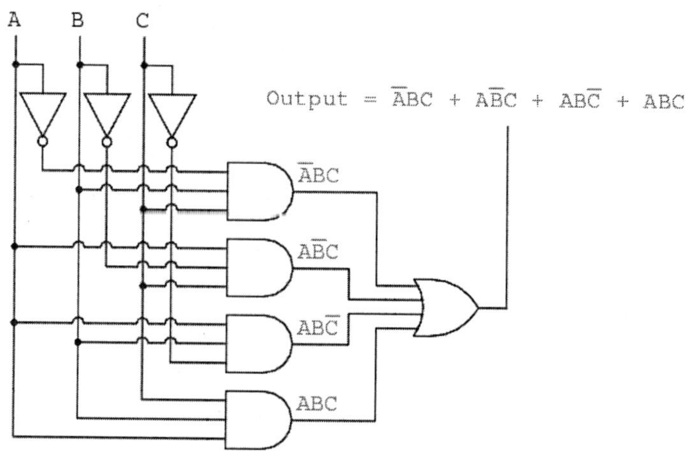

Unfortunately, both of these circuits are quite complex, and could benefit from simplification. Using Boolean algebra techniques, the expression may be significantly simplified:

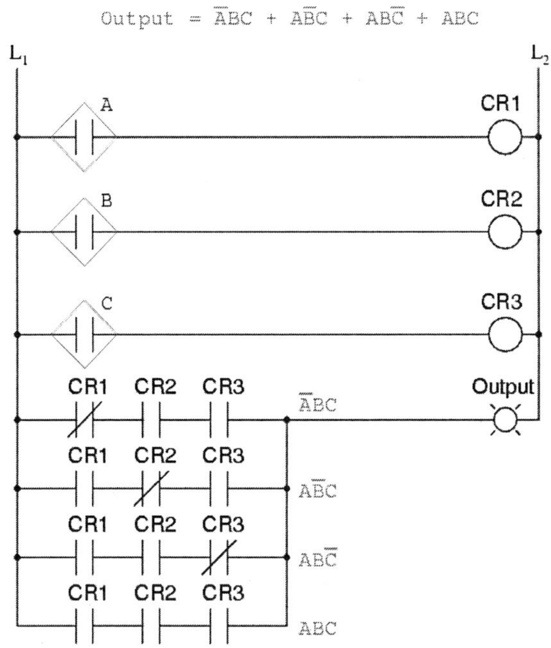

As a result of the simplification, we can now build much simpler logic circuits performing the same function, in either gate or relay form:

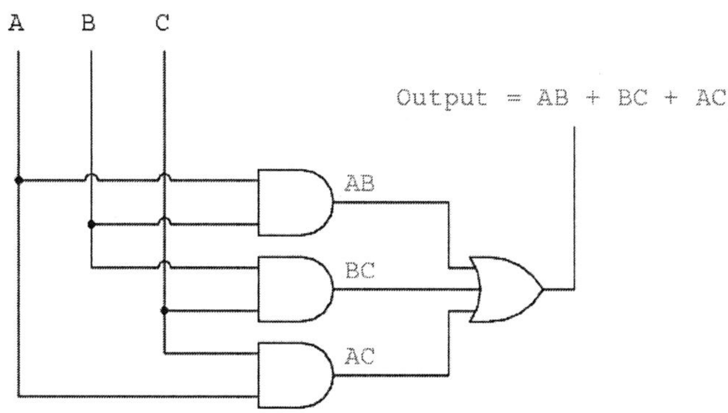

Either one of these circuits will adequately perform the task of operating the incinerator waste valve based on a flame verification from two out of the three flame sensors. At minimum, this is what we need to have a safe incinerator system.

We can, however, extend the functionality of the system by adding to it logic circuitry designed to detect if any one of the sensors does not agree with the other two.

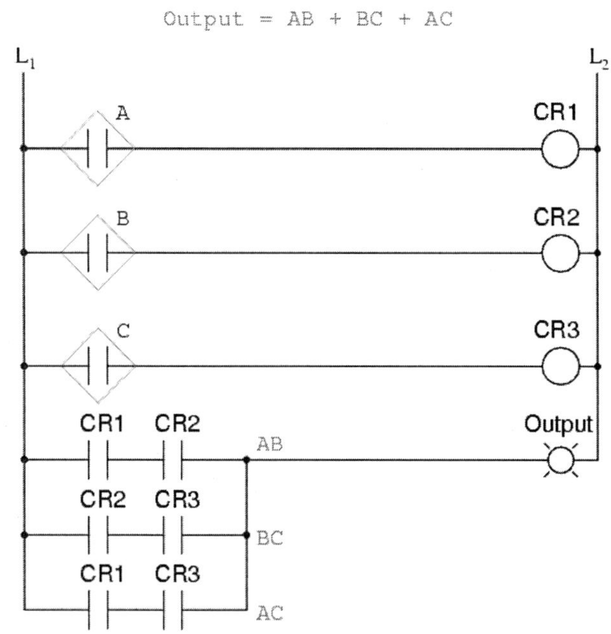

If all three sensors are operating properly, they should detect flame with equal accuracy. Thus, they should either all register "low" (000: no flame) or all register "high" (111: good flame). Any other output combination (001, 010, 011, 100, 101, or 110) constitutes a disagreement between sensors, and may therefore serve as an indicator of a potential sensor failure. If we added circuitry to detect any one of the six "sensor disagreement" conditions, we could use the output of that circuitry to activate an alarm.

Whoever is monitoring the incinerator would then exercise judgment in either continuing to operate with a possible failed sensor (inputs: 011, 101, or 110), or shut the incinerator down to be absolutely safe. Also, if the incinerator is shut down (no flame), and one or more of the sensors still indicates flame (001, 010, 011, 100, 101, or 110) while the other(s) indicate(s) no flame, it will be known that a definite sensor problem exists.

The first step in designing this "sensor disagreement" detection circuit is to write a truth table describing its behaviour. Since we already have a truth table describing the output of the "good flame" logic circuit, we can simply add another output column to the table to represent the second circuit, and make a table representing the entire logic system:

Output = 0 (close valve) Output = 1 (open valve)

Output = 0 (sensors agree) Output = 1 (sensors disagree)

sensor inputs			Good flame	Sensor disagreement
A	B	C	Output	Output
0	0	0	0	0
0	0	1	0	1
0	1	0	0	1
0	1	1	1	1
1	0	0	0	1
1	0	1	1	1
1	1	0	1	1
1	1	1	1	0

While it is possible to generate a Sum-Of-Products expression for this new truth table column, it would require six terms, of three variables each! Such a Boolean expression would require many steps to simplify, with a large potential for making algebraic errors:

Output = 0 (close valve) Output = 1 (open valve)

Output = 0 (sensors agree) Output = 1 (sensors disagree)

sensor inputs			Good flame	Sensor disagreement	
A	B	C	Output	Output	
0	0	0	0	0	
0	0	1	0	1	$\bar{A}\bar{B}C$
0	1	0	0	1	$\bar{A}B\bar{C}$
0	1	1	1	1	$\bar{A}BC$
1	0	0	0	1	$A\bar{B}\bar{C}$
1	0	1	1	1	$A\bar{B}C$
1	1	0	1	1	$AB\bar{C}$
1	1	1	1	0	

$$\text{Output} = \bar{A}\bar{B}C + \bar{A}B\bar{C} + \bar{A}BC + A\bar{B}\bar{C} + A\bar{B}C + AB\bar{C}$$

An alternative to generating a Sum-Of-Products expression to account for all the "high" (1) output conditions in the truth table is to generate a *Product-Of-Sums*, or *POS*, expression, to account for all the "low" (0) output conditions instead.

Being that there are much fewer instances of a "low" output in the last truth table column, the resulting Product-Of-Sums expression should contain fewer terms. As its name suggests, a Product-Of-Sums expression is a set of added terms (*sums*), which are multiplied (*product*) together. An example of a POS expression would be (A + B)(C + D), the product of the sums "A + B" and "C + D".

To begin, we identify which rows in the last truth table column have "low" (0) outputs, and write a Boolean sum term that would equal 0 for that row's input conditions. For instance, in the first row of the truth table, where A=0, B=0, and C=0, the sum term would be (A + B + C), since that term would have a value of 0 if and only if A=0, B=0, and C=0:

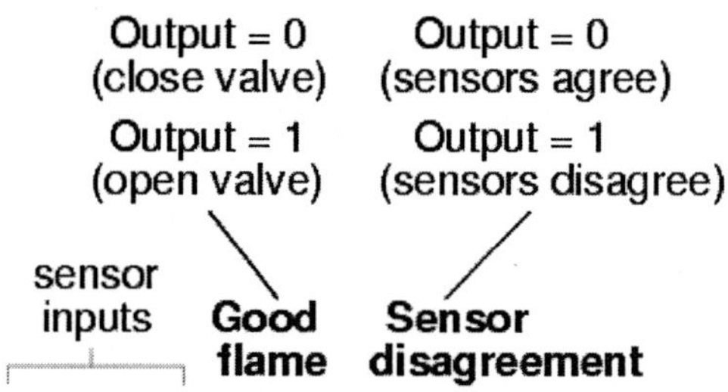

A	B	C	Output	Output	
0	0	0	0	0	(A + B + C)
0	0	1	0	1	
0	1	0	0	1	
0	1	1	1	1	
1	0	0	0	1	
1	0	1	1	1	
1	1	0	1	1	
1	1	1	1	0	

Only one other row in the last truth table column has a "low" (0) output, so all we need is one more sum term to complete our Product-Of-Sums expression. This last sum term represents a 0 output for an input condition of A=1, B=1 and C=1. Therefore, the term must be written as (A' + B'+ C'), because only the sum of the *complemented* input variables would equal 0 for that condition only:

Output = 0 (close valve) | Output = 0 (sensors agree)
Output = 1 (open valve) | Output = 1 (sensors disagree)

sensor inputs | **Good flame** | **Sensor disagreement**

A	B	C	Output	Output	
0	0	0	0	0	$(A + B + C)$
0	0	1	0	1	
0	1	0	0	1	
0	1	1	1	1	
1	0	0	0	1	
1	0	1	1	1	
1	1	0	1	1	
1	1	1	1	0	$(\overline{A} + \overline{B} + \overline{C})$

The completed Product-Of-Sums expression, of course, is the multiplicative combination of these two sum terms:

Output = 0 (close valve) | Output = 0 (sensors agree)
Output = 1 (open valve) | Output = 1 (sensors disagree)

sensor inputs | **Good flame** | **Sensor disagreement**

A	B	C	Output	Output	
0	0	0	0	0	$(A + B + C)$
0	0	1	0	1	
0	1	0	0	1	
0	1	1	1	1	
1	0	0	0	1	
1	0	1	1	1	
1	1	0	1	1	
1	1	1	1	0	$(\overline{A} + \overline{B} + \overline{C})$

$$\text{Output} = (A + B + C)(\overline{A} + \overline{B} + \overline{C})$$

Whereas a Sum-Of-Products expression could be implemented in the form of a set of AND gates with their outputs connecting to a single OR gate, a Product-Of-Sums expression can be implemented as a set of OR gates feeding into a single AND gate:

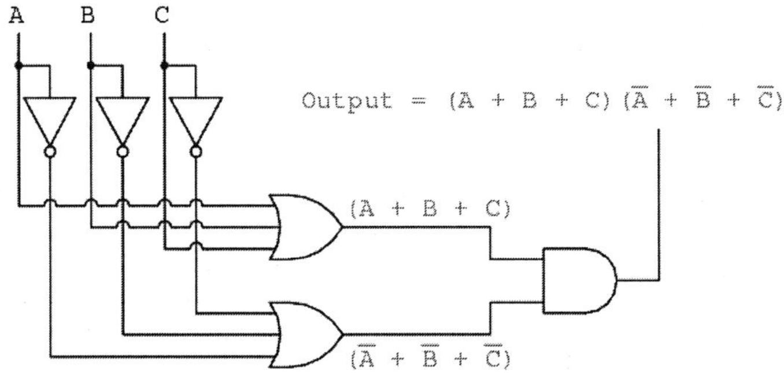

Correspondingly, whereas a Sum-Of-Products expression could be implemented as a parallel collection of series-connected relay contacts, a Product-Of-Sums expression can be implemented as a series collection of parallel-connected relay contacts:

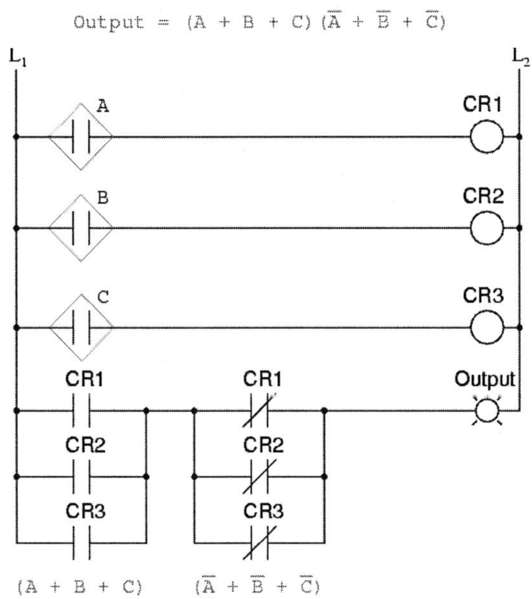

The previous two circuits represent different versions of the "sensor disagreement" logic circuit only, not the "good flame" detection circuit(s). The entire logic system would be the combination of both "good flame" and "sensor disagreement" circuits, shown on the same diagram.

Implemented in a Programmable Logic Controller (PLC), the entire logic system might resemble something like this:

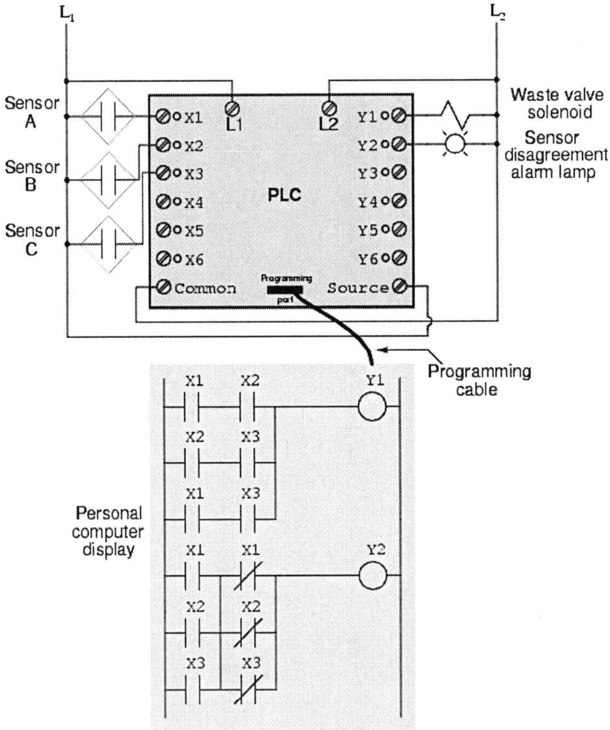

As you can see, both the Sum-Of-Products and Products-Of-Sums standard Boolean forms are powerful tools when applied to truth tables. They allow us to derive a Boolean expression—and ultimately, an actual logic circuit—from nothing but a truth table, which is a written specification for what we want a logic circuit to do. To be able to go from a written specification to an actual circuit using simple, deterministic procedures means that it is possible to automate the design process for a digital circuit. In other words, a computer could be programmed to design a custom logic circuit from a truth table specification! The steps to take from a truth table to the final circuit are so unambiguous and direct that it requires little, if any, creativity or other original thought to execute them.

Review:

Sum-Of-Products, or *SOP*, Boolean expressions may be generated from truth tables quite easily, by determining which rows of the table have an output of 1, writing one product term for each row, and finally summing all the product terms. This creates a Boolean expression representing the truth table as a whole.

Sum-Of-Products expressions lend themselves well to implementation as a set of AND gates (products) feeding into a single OR gate (sum).

Product-Of-Sums, or *POS*, Boolean expressions may also be generated from truth tables quite easily, by determining which rows of the table have an output of 0, writing one sum term for each row, and finally multiplying all the sum terms. This creates a Boolean expression representing the truth table as a whole. Product-Of-Sums expressions lend themselves well to implementation as a set of OR gates (sums) feeding into a single AND gate (product).

Logical Boolean Expression

George Boole believed in what he called the 'process of analysis', that is, the process by which combinations of interpretable symbols are obtained.

It is the use of these symbols according to well-determined methods of combination that he believed presented 'true calculus'. Today, all our computers use Boole's logic system - using microchips that contain thousands of tiny electronic switches arranged into logical 'gates' that produce predictable and reliable conclusions.

The basic logic gates are AND, OR and NOT. It is these gates, used in different combinations, that allow the computer to execute its operations using binary language.

Each gate assesses information (consisting of high or low voltages) in accordance with predetermined rules, and produces a single high or low voltage logical conclusion. The voltage itself represents the binary yes-no, true-false, one-zero concept.

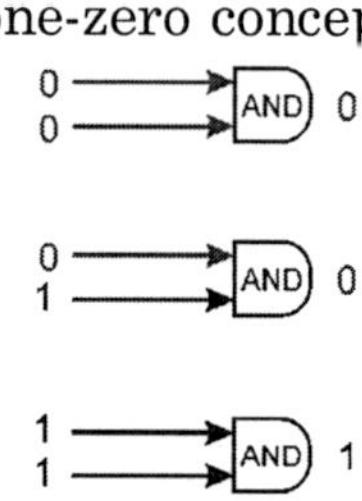

AND gates will only yield a TRUE result (that is, a binary 1) if all input is TRUE. Therefore, the top two gates will produce a FALSE (binary 0) result.

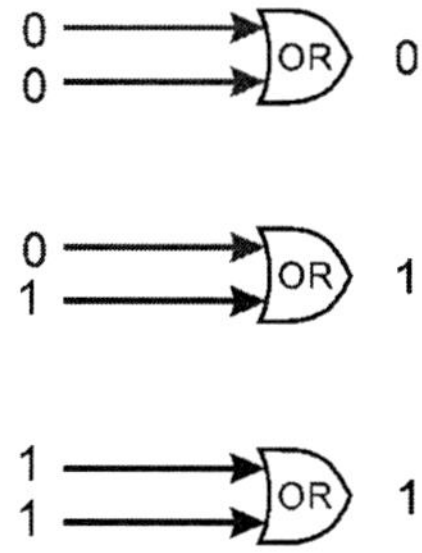

OR gates are less fussy. An OR gate will send a TRUE result if any of its input is TRUE. Only if every input is FALSE will it produce a 0 result.

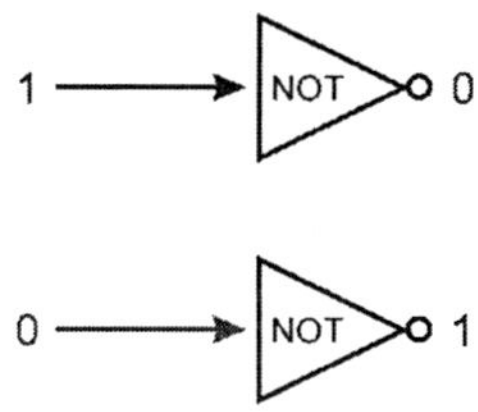

NOT gates, on the other hand, do just one thing. A NOTgate accepts only a single input, either TRUE or FALSE, which it promptly reverses. NOT gates can be combined with AND or OR to give NAND (not and) orNOR (not or) gates. These new gates process input in the usual manner and then reverse the result.

Putting them all together produces something like this ('this', by the way, is known as a half-adder). The top gate inputs 1 and 0 through OR to give a 1 result. The bottom gate channels 1 and 0 through AND to produce 0, which is inverted through the NOTgate into a second 1. The last AND gate sees two inputs of 1 and promptly spits out a TRUE (1) result. Thus, 1 + 0 = 1.

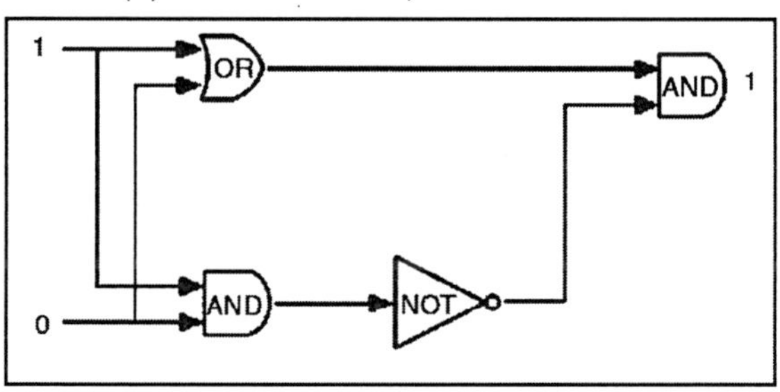

By adjusting the gates, binary adders can be used not only for addition, but for subtraction, multiplication, and division as well, enabling them to control the computer's data flow and to process instructions.

Chapter 6

System of Linear Equations

A system of linear equations (or linear system) is a collection of linear equations involving the same set of variables. For example,

$$3x - 2y - z = 1$$

$$2x - 2y + 4z = -2$$

$$-x \frac{1}{2} y - z = 0$$

is a system of three equations in the three variables *x, y, z*. A solution to a linear system is an assignment of numbers to the variables such that all the equations are simultaneously satisfied. A solution to the system above is given by

$$x = 1$$

$$y = -2$$

$$z = -2$$

since it makes all three equations valid.

In mathematics, the theory of linear systems is a branch of linear algebra, a subject which is fundamental to modern mathematics. Computational algorithms for finding the solutions are an important part of numerical linear algebra, and such methods play a prominent role in engineering, physics, chemistry, computer science, and economics. A system of non-linear equations can often be approximated by a linear system, a helpful technique when making a mathematical model or computer simulation of a relatively complex system.

Elementary Example

The simplest kind of linear system involves two equations and two variables:

$$2x+3y=6$$
$$4x|9y=15.$$

One method for solving such a system is as follows. First, solve the top equation for x in terms of y:

$$x=3-\frac{3}{2}y.$$

Now substitute this expression for x into the bottom equation:

$$4\left(3-\frac{3}{2}y\right)+9y=15.$$

This results in a single equation involving only the variable y. Solving gives $y = 1$, and substituting this back into the equation for x yields $x = 3/2$. This method generalizes to systems with additional variables.

General Form

A general system of m linear equations with n unknowns can be written as

$$\begin{array}{cccc} a_{11}x_1 + & a_{12}x_2 + \cdots + & a_{1n}x_n = & b_1 \\ a_{21}x_1 + & a_{22}x_2 + \cdots + & a_{2n}x_n = & b_2 \\ \vdots & \vdots & \vdots & \vdots \\ a_{m1}x_1 + & a_{m2}x_2 + \cdots + & a_{mn}x_n = & b_{m.} \end{array}$$

Here $x_1, x_2, ..., x_n$ are the unknowns, $a_{11}, a_{12}, ..., a_{mn}$ are the coefficients of the system, and $b_1, b_2, ..., b_m$ are the constant terms.

Often the coefficients and unknowns are real or complex numbers, but integers and rational numbers are also seen, as are polynomials and elements of an abstract algebraic structure.

Vector Equation

One extremely helpful view is that each unknown is a weight for a column vector in a linear combination.

$$x_1\begin{bmatrix} a_{11} \\ a_{21} \\ \vdots \\ a_{m1} \end{bmatrix} + x_2\begin{bmatrix} a_{12} \\ a_{22} \\ \vdots \\ a_{m2} \end{bmatrix} + \cdots + x_n\begin{bmatrix} a_{1n} \\ a_{2n} \\ \vdots \\ a_{mn} \end{bmatrix} = \begin{bmatrix} b_1 \\ b_2 \\ \vdots \\ b_m \end{bmatrix}$$

This allows all the language and theory of *vector spaces* (or more generally, *modules*) to be brought to bear. For example, the collection of all possible linear combinations of the vectors on the left-hand side is called their *span*, and the equations have a solution just when the right-hand vector is within that span. If every vector within that span has exactly one expression as a linear combination of the given left-hand vectors, then any solution is unique.

In any event, the span has a *basis* of linearly independent vectors that do guarantee exactly one expression; and the number of vectors in that basis (its *dimension*) cannot be larger than m or n, but it can be smaller. This is important because if we have m independent vectors a solution is guaranteed regardless of the right-hand side, and otherwise not guaranteed.

Matrix Equation

The vector equation is equivalent to a matrix equation of the form

$$Ax = b$$

where A is an $m \times n$ matrix, x is a column vector with n entries, and b is a column vector with m entries.

$$A = \begin{bmatrix} a_{11} & a_{12} & \cdots & a_{1n} \\ a_{21} & a_{22} & \cdots & a_{2n} \\ \vdots & \vdots & \ddots & \vdots \\ a_{m1} & a_{m2} & \cdots & a_{mn} \end{bmatrix}, \quad \mathbf{x} = \begin{bmatrix} x_1 \\ x_2 \\ \vdots \\ x_n \end{bmatrix}, \quad \mathbf{b} = \begin{bmatrix} b_1 \\ b_2 \\ \vdots \\ b_m \end{bmatrix}$$

The number of vectors in a basis for the span is now expressed as the *rank* of the matrix.

Solution Set

The solution set for the equations $x - y = -1$ and $3x + y = 9$ is the single point (2, 3). A solution of a linear system is an assignment of values to the variables $x_1, x_2, \ldots, x_n$ such that each of the equations is satisfied. The set of all possible solutions is called the solution set.

A linear system may behave in any one of three possible ways:

1. The system has *infinitely many solutions.*
2. The system has a single *unique solution.*
3. The system has no *solution.*

Geometric Interpretation

For a system involving two variables (x and y), each linear equation determines a line on the xy-plane. Because a solution to a linear system must satisfy all of the equations, the solution set is the intersection of these lines, and is hence either a line, a single point, or the empty set.

For three variables, each linear equation determines a plane in three-dimensional space, and the solution set is the intersection of these planes. Thus the solution set may be a plane, a line, a single point, or the empty set.

For n variables, each linear equations determines a hyperplane in n-dimensional space. The solution set is the intersection of these hyperplanes, which may be a flat of any dimension.

General Behaviour

The solution set for two equations in three variables is usually a line.

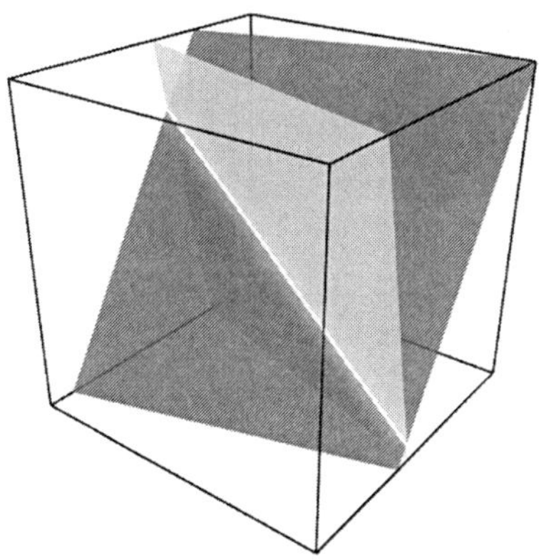

In general, the behaviour of a linear system is determined by the relationship between the number of equations and the number of unknowns:

1. Usually, a system with fewer equations than unknowns has infinitely many solutions. Such a system is also known as an underdetermined system.
2. Usually, a system with the same number of equations and unknowns has a single unique solution.
3. Usually, a system with more equations than unknowns has no solution.

In the first case, the dimension of the solution set is usually equal to $n - m$, where n is the number of variables and m is the number of equations.

The following pictures illustrate this trichotomy in the case of two variables:

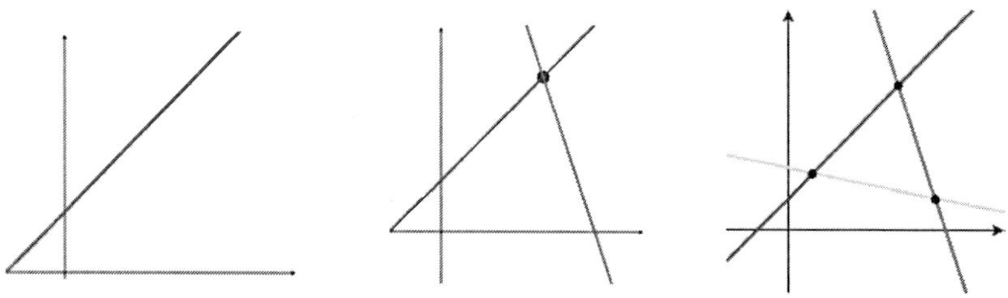

The first system has infinitely many solutions, namely all of the points on the blue line. The second system has a single unique solution, namely the intersection of the two lines. The third system has no solutions, since the three lines share no common point.

Keep in mind that the pictures above show only the most common case. It is possible for a system of two equations and two unknowns to have no solution (if the two lines are parallel), or for a system of three equations and two unknowns to be solvable (if the three lines intersect at a single point). In general, a system of linear equations may behave differently than expected if the equations are linearly dependent, or if two or more of the equations are inconsistent.

Properties

Consistency: The equations $3x + 2y = 6$ and $3x + 2y = 12$ are inconsistent.

The equations of a linear system are consistent if they possess a common solution, and inconsistent otherwise. When the equations are inconsistent, it is possible to derive a contradiction from the equations, such as the statement that $0 = 1$.

For Example, the Equations

$$3x+2y=6 \quad \text{and} \quad 3x+2y=12$$

are inconsistent. In attempting to find a solution, we tacitly assume that there is a solution; that is, we assume that the value of x in the first equation must be the same as the value of x in the second equation (the same is assumed to simultaneously be true for the value of y in both equations). Applying the substitution property (for 3x+2y) yields the equation $6 = 12$, which is a false statement. This therefore contradicts our assumption that the system had a solution and we conclude that our assumption was false; that is, the system in fact has no solution. The graphs of these equations on the xy-plane are a pair of parallel lines. It is possible for three linear equations to be inconsistent, even though any two of the equations are consistent together. For example, the equations

$$x + y = 1$$

$$2x + y = 1$$

$$3x + 2y = 3$$

are inconsistent. Adding the first two equations together gives $3x + 2y = 2$, which can be subtracted from the third equation to yield $0 = 1$. Note that any two of these equations have a common solution. The same phenomenon can occur for any number of equations. In general, inconsistencies occur if the left-hand sides of the equations in a system are linearly dependent, and the constant terms do not satisfy the dependence relation. A system of equations whose left-hand sides are linearly independent is always consistent.

Independence

The equations of a linear system are independent if none of the equations can be derived algebraically from the others. When the equations are independent, each equation contains new information about the variables, and removing any of the equations increases the size of the solution set. For linear equations, logical independence is the same as linear independence. The equations $x - 2y = -1$, $3x + 5y = 8$, and $4x + 3y = 7$ are not linearly independent.

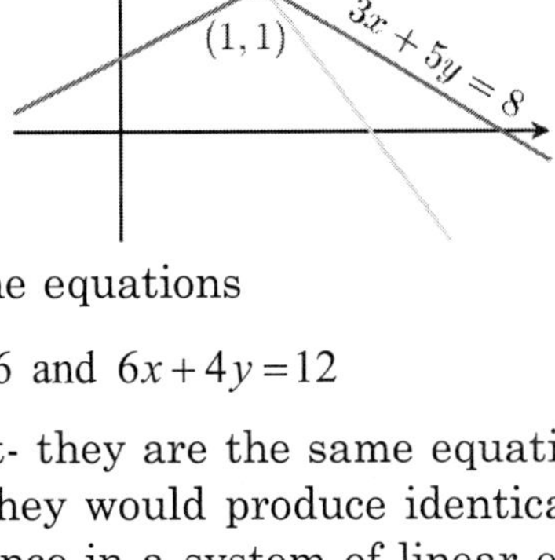

For example, the equations

$$3x + 2y = 6 \text{ and } 6x + 4y = 12$$

are not independent- they are the same equation when scaled by a factor of two, and they would produce identical graphs. This is an example of equivalence in a system of linear equations.

For a more complicated example, the equations

$$x - 2y = -1$$
$$3x + 5y = 8$$
$$4x + 3y = 7$$

are not independent, because the third equation is the sum of the other two. Indeed, any one of these equations can be derived from the other two, and any one of the equations can be removed without affecting the solution set. The graphs of these equations are three lines that intersect at a single point.

Equivalence

Two linear systems using the same set of variables are equivalent if each of the equations in the second system can be derived algebraically from the equations in the first system, and vice-versa. Equivalent systems convey precisely the same information about the values of the variables. In particular, two linear systems are equivalent if and only if they have the same solution set.

Solving a Linear System

There are several algorithms for solving a system of linear equations.

Describing the Solution

When the solution set is finite, it is usually described in set notation. For example, the solution set 2, 3, and 4 would be written: {2,3,4}

It can be difficult to describe a set with infinite solutions. Typically, some of the variables are designated as free (or independent, or as parameters), meaning that they are allowed to take any value, while the remaining variables are dependent on the values of the free variables.

For example, consider the following system:

$$x + 3y - 2z = 5$$
$$3x + 5y + 6z = 7$$

The solution set to this system can be described by the following equations:

$$x = -7z - 1 \text{ and } y = 3z + 2.$$

Here z is the free variable, while x and y are dependent on z. Any point in the solution set can be obtained by first choosing a value for z, and then computing the corresponding values for x and y.

Each free variable gives the solution space one degree of freedom, the number of which is equal to the dimension of the solution set.

For example, the solution set for the above equation is a line, since a point in the solution set can be chosen by specifying the value of the parameter z. An infinite solution of higher order may describe a plane, or higher dimensional set.

Different choices for the free variables may lead to different descriptions of the same solution set. For example, the solution to the above equations can alternatively be described as follows:

$$y = -\frac{3}{7}x + \frac{11}{7} \text{ and } z = -\frac{1}{7}x - \frac{1}{7}.$$

Here x is the free variable, and y and z are dependent.

Elimination of Variables

The simplest method for solving a system of linear equations is to repeatedly eliminate variables. This method can be described as follows:

1. In the first equation, solve for the one of the variables in terms of the others.
2. Plug this expression into the remaining equations. This yields a system of equations with one fewer equation and one fewer unknown.
3. Continue until you have reduced the system to a single linear equation.
4. Solve this equation, and then back-substitute until the entire solution is found.

For example, consider the following system:

$$\begin{aligned} x + 3y - 2z &= 5 \\ 3x + 5y + 6z &= 7 \\ 2x + 4y + 3z &= 8 \end{aligned}$$

Solving the first equation for x gives $x = 5 + 2z - 3y$, and plugging this into the second and third equation yields

$$-4y + 12z = -8$$
$$-2y + 7z = -2$$

Solving the first of these equations for y yields $y = 2 + 3z$, and plugging this into the third equation yields $z = 2$. We now have:

$$x = 5 + 2z - 3y$$
$$y = 2 + 3z$$
$$z = 2$$

Substituting $z = 2$ into the second equation gives $y = 8$, and substituting $z = 2$ and $y = 8$ into the first equation yields $x = -15$. Therefore, the solution set is the single point $(x, y, z) = (-15, 8, 2)$.

Row Reduction

In row reduction, the linear system is represented as an augmented matrix:

$$\left[\begin{array}{ccc|c} 1 & 3 & -2 & 5 \\ 3 & 5 & 6 & 7 \\ 2 & 4 & 3 & 8 \end{array}\right].$$

This matrix is then modified using elementary row operations until it reaches reduced row echelon form. There are three types of elementary row operations:

Type 1: Swap the positions of two rows.

Type 2: Multiply a row by a nonzero scalar.

Type 3: Add to one row a scalar multiple of another.

Because these operations are reversible, the augmented matrix produced always represents a linear system that is equivalent to the original.

There are several specific algorithms to row-reduce an augmented matrix, the simplest of which are Gaussian elimination and Gauss-Jordan elimination. The following computation shows Gaussian elimination applied to the matrix above:

The last matrix is in reduced row echelon form, and represents the system $x = -15$, $y = 8$, $z = 2$. A comparison with the example in the previous section on the algebraic elimination of variables shows that these two methods are in fact the same; the difference lies in how the computations are written down.

$$\begin{bmatrix} 1 & 3 & -2 & 5 \\ 3 & 5 & 6 & 7 \\ 2 & 4 & 3 & 8 \end{bmatrix} \sim \begin{bmatrix} 1 & 3 & -2 & 5 \\ 0 & -4 & 12 & -8 \\ 2 & 4 & 3 & 8 \end{bmatrix} \sim \begin{bmatrix} 1 & 3 & -2 & 5 \\ 0 & -4 & 12 & -8 \\ 0 & -2 & 7 & -2 \end{bmatrix} \sim \begin{bmatrix} 1 & 3 & -2 & 5 \\ 0 & 1 & -3 & 2 \\ 0 & -2 & 7 & -2 \end{bmatrix} \sim$$

$$\begin{bmatrix} 1 & 3 & -2 & 5 \\ 0 & 1 & -3 & 2 \\ 0 & 0 & 1 & 2 \end{bmatrix} \sim \begin{bmatrix} 1 & 3 & -2 & 5 \\ 0 & 1 & 0 & 8 \\ 0 & 0 & 1 & 2 \end{bmatrix} \sim \begin{bmatrix} 1 & 3 & 0 & 9 \\ 0 & 1 & 0 & 8 \\ 0 & 0 & 1 & 2 \end{bmatrix} \sim \begin{bmatrix} 1 & 0 & 0 & -15 \\ 0 & 1 & 0 & 8 \\ 0 & 0 & 1 & 2 \end{bmatrix}.$$

Cramer's Rule

Cramer's rule is an explicit formula for the solution of a system of linear equations, with each variable given by a quotient of two determinants. For example, the solution to the system

$$\begin{aligned} x+3y-2z &= 5 \\ 3x+5y+6z &= 7 \\ 2x+4y+3z &= 8 \end{aligned}$$

is given by

$$x = \frac{\begin{vmatrix} 5 & 3 & -2 \\ 7 & 5 & 6 \\ 8 & 4 & 3 \end{vmatrix}}{\begin{vmatrix} 1 & 3 & -2 \\ 3 & 5 & 6 \\ 2 & 4 & 3 \end{vmatrix}}, \quad y = \frac{\begin{vmatrix} 1 & 5 & -2 \\ 3 & 7 & 6 \\ 2 & 8 & 3 \end{vmatrix}}{\begin{vmatrix} 1 & 3 & -2 \\ 3 & 5 & 6 \\ 2 & 4 & 3 \end{vmatrix}}, \quad z = \frac{\begin{vmatrix} 1 & 3 & 5 \\ 3 & 5 & 7 \\ 2 & 4 & 8 \end{vmatrix}}{\begin{vmatrix} 1 & 3 & -2 \\ 3 & 5 & 6 \\ 2 & 4 & 3 \end{vmatrix}}.$$

For each variable, the denominator is the determinant of the matrix of coefficients, while the numerator is the determinant of a matrix in which one column has been replaced by the vector of constant terms.

Though Cramer's rule is important theoretically, it has little practical value for large matrices, since the computation of large determinants is somewhat cumbersome. (Indeed, large determinants are most easily computed using row reduction.) Further, Cramer's rule has very poor numerical properties, making it unsuitable for solving even small systems reliably, unless the operations are performed in rational arithmetic with unbounded precision.

Other Methods

While systems of three or four equations can be readily solved by hand, computers are often used for larger systems. The standard algorithm for solving a system of linear equations is based on Gaussian elimination with some modifications. Firstly, it is essential to avoid division by small numbers, which may lead to inaccurate results. This can be done by reordering the equations if necessary, a process known as *pivoting*.

Secondly, the algorithm does not exactly do Gaussian elimination, but it computes the LU decomposition of the matrix A. This is mostly an organizational tool, but it is much quicker if one has to solve several systems with the same matrix A but different vectors b.

If the matrix A has some special structure, this can be exploited to obtain faster or more accurate algorithms. For instance, systems with a symmetric positive definite can be solved twice as fast with the Cholesky decomposition. Levinson recursion is a fast method for Toeplitz matrices. Special methods exist also for matrices with many zero elements (so-called sparse matrices), which appear often in applications.

A completely different approach is often taken for very large systems, which would otherwise take too much time or memory. The idea is to start with an initial approximation to the solution (which does not have to be accurate at all), and to change this approximation in several steps to bring it closer to the true solution. Once the approximation is sufficiently accurate, this is taken to be the solution to the system. This leads to the class of iterative methods.

Homogeneous Systems

A system of linear equations is homogeneous if all of the constant terms are zero:

$$\begin{array}{cccc} a_{11}x_1 + & a_{12}x_2 + \cdots + & a_{1n}x_n = & 0 \\ a_{21}x_1 + & a_{22}x_2 + \cdots + & a_{2n}x_n = & 0 \\ \vdots & \vdots & \vdots & \vdots \\ a_{m1}x_1 + & a_{m2}x_2 + \cdots + & a_{mn}x_n = & 0. \end{array}$$

A homogeneous system is equivalent to a matrix equation of the form

$$Ax = 0$$

where A is an $m \times n$ matrix, x is a column vector with n entries, and 0 is the zero vector with m entries.

Solution Set

Every homogeneous system has at least one solution, known as the zero solution (or trivial solution), which is obtained by assigning the value of zero to each of the variables. The solution set has the following additional properties:

1. If u and v are two vectors representing solutions to a homogeneous system, then the vector sum u + v is also a solution to the system.
2. If u is a vector representing a solution to a homogeneous system, and r is any scalar, then ru is also a solution to the system.

These are exactly the properties required for the solution set to be a linear subspace of R^n.

In particular, the solution set to a homogeneous system is the same as the null space of the corresponding matrix A.

Relation to Nonhomogeneous Systems

There is a close relationship between the solutions to a linear system and the solutions to the corresponding homogeneous system:

$Ax = b$ and $Ax = 0$.

Specifically, if p is any specific solution to the linear system Ax = b, then the entire solution set can be described as

$\{p + v : v \text{ is any solution to } Ax = 0\}$.

Geometrically, this says that the solution set for Ax = b is a translation of the solution set for Ax = 0. Specifically, the flat for the first system can be obtained by translating the linear subspace for the homogeneous system by the vector p. This reasoning only applies if the system Ax = b has at least one solution. This occurs if and only if the vector b lies in the image of the linear transformation A.

Matrix Solution of Linear Systems

A matrix (plural: matrices) is a rectangular array of numbers arranged in rows and columns and enclosed in brackets. Each number in the matrix is called an element of the matrix.

Each of the following is an example of a matrix:

$$\begin{bmatrix} 5 & -1 & 0 \\ -2 & 6 & 9 \end{bmatrix} \begin{bmatrix} 4 & -3 \\ -7 & 10 \end{bmatrix}$$

We are going to look at a technique for solving linear systems of equations using matrices.

A matrix derived from a linear system of equations, each in standard form, is called the augmented matrix of the system. The augmented matrix for a linear system is a matrix with the system's variables and equal signs eliminated. For example:

System:

$$x - 2y - z = 2$$
$$2x - y = 4$$
$$-x + y - 2z = -4$$

Augmented Matrix:

$$\begin{bmatrix} 1 & -2 & -1 & 2 \\ 2 & -1 & 0 & 4 \\ -1 & 1 & -2 & -4 \end{bmatrix}$$

Each row of the augmented matrix represents one equation of the system. In each row, the coefficient of x is in the first column, the coefficient of y is in the second column, the coefficient of z is in the third column, and the constant is in the fourth column.

Notice that if there is a missing term, as the "z" term in the second equation, a zero represents its coefficient in the augmented matrix.

Matrix Row Transformations

For any augmented matrix of a system of linear equations, the following row transformations will result in the matrix of an equivalent system:

1. Any two rows may be interchanged.
2. The elements of any row may be multiplied (or divided) by the same nonzero number.
3. Any row may be changed by adding to its elements a multiple of the corresponding elements of another row.

To solve a system of equations using the augmented matrix, we will use matrix row transformations to convert the augmented matrix into triangular form.

A matrix is in triangular form if there are 1's down the main diagonal and 0's below the diagonal.

The following matrix is in triangular form:

$$\begin{bmatrix} 1 & A & B & C \\ 0 & 1 & D & E \\ 0 & 0 & 1 & F \end{bmatrix}$$

The system of equations represented by this matrix is:

x + Ay + Bz = C

y + Dz = E

z = F

The system is easily solved from this form.

To solve a linear system using matrix row transformations, we begin with the augmented matrix. We then use row transformations to obtain an equivalent matrix with 1's down the diagonal and 0's below the diagonal. We will proceed column by column from left to right.

In each column, we will start by obtaining a 1 in the diagonal position; then obtain 0's below the 1.

This method is called the Gauss-Jordan method.

Example: Solve:

x + y – z = -2

2x – y + z = 5

-x + 2y + 2z = 1

Solution:

$$\begin{bmatrix} 1 & 1 & -1 & -2 \\ 2 & -1 & 1 & 5 \\ -1 & 2 & 2 & 1 \end{bmatrix}$$

We begin by writing the system as an augmented matrix

We already have a 1 in the diagonal position of first column. Now we want 0's below the 1. The first 0 can be obtained by multiplying row 1 by -2 and adding the results to row 2:

$$\begin{bmatrix} 1 & 1 & -1 & -2 \\ 0 & -3 & 3 & 9 \\ -1 & 2 & 2 & 1 \end{bmatrix}$$

Row 1 is unchanged:

-2 times row 1 is added to row 2.

Row 3 is unchanged.

The second 0 can be obtained by adding row 1 to row 3:

$$\begin{bmatrix} 1 & 1 & -1 & -2 \\ 0 & -3 & 3 & 9 \\ 0 & 3 & 1 & -1 \end{bmatrix}$$

Row 1 is unchanged.

Row 2 is unchanged.

Row 1 is added to Row 3.

Moving to the second column, we want a 1 in the diagonal position (where there is now –3). We get this by dividing every element in row 2 by -3:

$$\begin{bmatrix} 1 & 1 & -1 & -2 \\ 0 & 1 & -1 & -3 \\ 0 & 3 & 1 & -1 \end{bmatrix}$$

Row 1 is unchanged.

Row 2 is divided by –3.

Row 3 is unchanged.

To obtain a 0 below the 1, we multiply row 2 by -3 and add it to the third row:

$$\begin{bmatrix} 1 & 1 & -1 & -2 \\ 0 & 1 & -1 & -3 \\ 0 & 0 & 4 & 8 \end{bmatrix}$$

Row 1 is unchanged.

Row 2 is unchanged.

-3 times row 2 is added to row 3.

The last step is to obtain a 1 on the diagonal in column 3. We divide each element in Row 3 by 4 to accomplish this:

$$\begin{bmatrix} 1 & 1 & -1 & -2 \\ 0 & 1 & -1 & -3 \\ 0 & 0 & 1 & 2 \end{bmatrix}$$

Row 1 is unchanged.

Row 2 is unchanged.

Each element in Row 3 is divided by 4.

Converting back to a system of equations, we have

$$x + y - z = -2$$

$$y - z = -3$$

$$z = 2$$

We have z = 2. To find y, we back-substitute 2 for z in the second equation:

$$y - z = -3$$

$$y - 2 = -3$$

$$y = -1$$

Finally, we back substitute -1 for y and 2 for z in the first equation:

$$x + y - z = -2$$

$$x + (-1) - 2 = -2$$

$$x - 3 = -2$$

$$x = 1$$

The solution is the ordered triple (1, -1, 2).

Example. Solve:

$$3x - 4y + 4z = 7$$

$$x - y - 2z = 2$$

$$2x - 3y + 6z = 5$$

Solution: Start with the augmented matrix:

$$\begin{bmatrix} 3 & -4 & 4 & 7 \\ 1 & -1 & -2 & 2 \\ 2 & -3 & 6 & 5 \end{bmatrix}$$

To obtain a 1 in the top position of the first column, interchange rows 1 and 2:

$$\begin{bmatrix} 1 & -1 & -2 & 2 \\ 3 & -4 & 4 & 7 \\ 2 & -3 & 6 & 5 \end{bmatrix}$$

To get 0's below the 1: multiply row 1 by -3 and add to row 2; multiply row 1 by –2 and add to row 3.

$$\begin{bmatrix} 1 & -1 & -2 & 2 \\ 0 & -1 & 10 & 1 \\ 0 & -1 & 10 & 1 \end{bmatrix}$$

To get a 1 in the diagonal position in the second column, multiply row 2 by –1:

$$\begin{bmatrix} 1 & -1 & -2 & 2 \\ 0 & 1 & -10 & -1 \\ 0 & -1 & 10 & 1 \end{bmatrix}$$

Add row 2 to row 3:

$$\begin{bmatrix} 1 & -1 & -2 & 2 \\ 0 & 1 & -10 & -1 \\ 0 & 0 & 0 & 0 \end{bmatrix}$$

Row 3 represents the equation 0 = 0 which is always true. This indicates that the system is dependent and has infinitely many solutions.

Row 2 represents the equation $y - 10z = -1$.

Solving this equation for y gives us:

$$y = 10z - 1.$$

Row 1 represents the equation $x - y - 2z = 2$.

Substitute $10z - 1$ for y and solve for x:

$$x - (10z - 1) - 2z = 2$$

$$x - 10z + 1 - 2z = 2$$

$$x - 12z + 1 = 2$$

$$x = 12z + 1$$

Therefore, every ordered triple of the form $(12z + 1, 10z - 1, z)$ is a solution of the system.

Chapter 7

Differential and Integral Calculus

Derivative

In calculus, a branch of mathematics, the derivative is a measure of how a function changes as its input changes.

Loosely speaking, a derivative can be thought of as how much one quantity is changing in response to changes in some other quantity; for example, the derivative of the position of a moving object with respect to time is the object's instantaneous velocity.

The derivative of a function at a chosen input value describes the best linear approximation of the function near that input value. Informally, the derivative is the ratio of the infinitesimal change of the output over the infinitesimal change of the input producing that change of output.

For a real-valued function of a single real variable, the derivative at a point equals the slope of the tangent line to the graph of the function at that point.

In higher dimensions, the derivative of a function at a point is a linear transformation called the linearization. A closely related notion is the differential of a function.

The process of finding a derivative is called differentiation. The reverse process is called *antidifferentiation*.

The fundamental theorem of calculus states that antidifferentiation is the same as integration.

Differentiation and integration constitute the two fundamental operations in single-variable calculus.

Differentiation and the Derivative

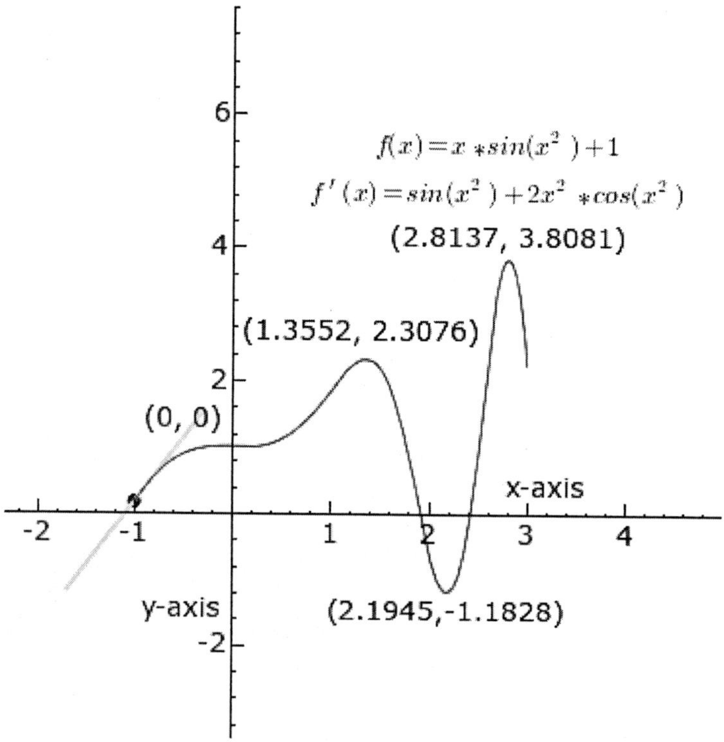

Figure: *At each point, the derivative of* $f(x) = 1 + x\sin x^2$ *is the slope of a line that is tangent to the curve. The line is always tangent to the blue curve; its slope is the derivative. Note derivative is positive where a green line appears, negative where a red line appears, and zero where a black line appears.*

Differentiation is a method to compute the rate at which a dependent output y changes with respect to the change in the independent input x. This rate of change is called the *derivative* of y with respect to x.

In more precise language, the dependence of y upon x means that y is a function of x. This functional relationship is often denoted $y = f(x)$, where f denotes the function. If x and y are real numbers, and if the graph of y is plotted against x, the derivative measures the slope of this graph at each point.

The simplest case is when y is a linear function of x, meaning that the graph of y divided by x is a straight line. In this case, $y = f(x) = m x + b$, for real numbers m and b, and the slope m is given by

$$m = \frac{\text{change in } y}{\text{change in } x} = \frac{\Delta y}{\Delta x},$$

where the symbol Δ (the uppercase form of the Greek letter Delta) is an abbreviation for "change in." This formula is true because

$$y + \Delta y = f(x+ \Delta x) = m\ (x + \Delta x) + b = m\ x + b + m\ \Delta x = y + m\Delta x.$$

It follows that $\Delta y = m\ \Delta x$.

This gives an exact value for the slope of a straight line. If the function f is not linear (i.e. its graph is not a straight line), however, then the change in y divided by the change in x varies: differentiation is a method to find an exact value for this rate of change at any given value of x.

Rate of Change as a Limiting Value

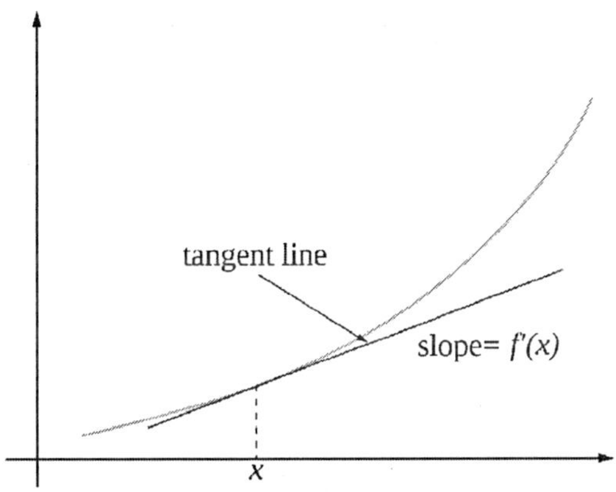

Figure: *The tangent line at (x, f(x))*

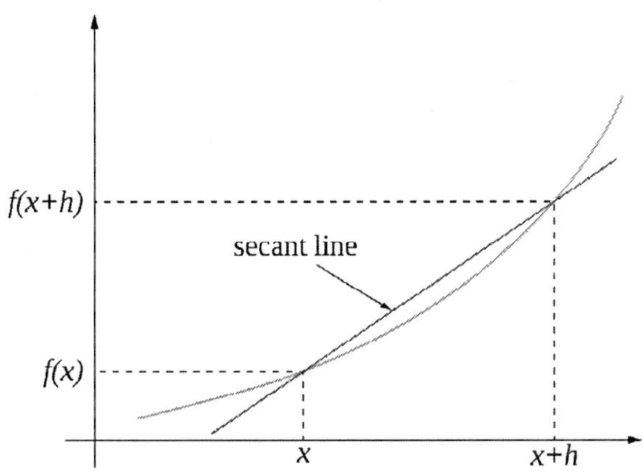

Figure: *The secant to curve y= f(x) determined by points (x, f(x)) and (x+h, f(x+h))*

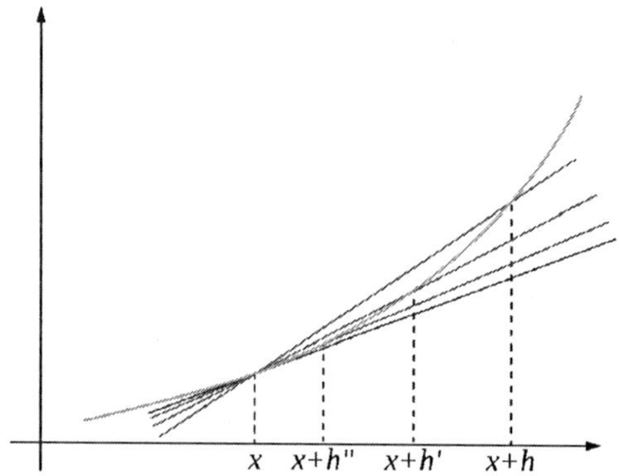

Figure: *The tangent line as limit of secants*

The idea is to compute the rate of change as the limiting value of the ratio of the differences Δy / Δx as Δx becomes infinitely small.

In Leibniz's notation, such an infinitesimal change in x is denoted by dx, and the derivative of y with respect to x is written

$$\frac{dy}{dx}$$

suggesting the ratio of two infinitesimal quantities. (The above expression is read as "the derivative of y with respect to x", "d y by d x", or "d y over d x". The oral form "d y d x" is often used conversationally, although it may lead to confusion.)

The most common approach to turn this intuitive idea into a precise definition uses limits, but there are other methods, such as non-standard analysis.

Definition via Difference Quotients

Let f be a real valued function. In classical geometry, the tangent line to the graph of the function f at a real number a was the unique line through the point (a, $f(a)$) that did *not* meet the graph of f transversally, meaning that the line did not pass straight through the graph. The derivative of y with respect to x at a is, geometrically, the slope of the tangent line to the graph of f at a. The slope of the tangent line is very close to the slope of the line through (a, $f(a)$) and a nearby point on the graph, for example ($a + h$, $f(a + h)$). These lines are called secant lines. A value of h close to zero gives a good approximation to the slope of the tangent line, and smaller values (in absolute value) of h will, in general, give better approximations. The slope m of the secant line is the difference between the y values of these points divided by the difference between the x values, that is,

$$m = \frac{\Delta f(a)}{\Delta a} = \frac{f(a+h)-f(a)}{(a+h)-(a)} = \frac{f(a+h)-f(a)}{h}.$$

This expression is Newton's difference quotient. The derivative is the value of the difference quotient as the secant lines approach the tangent line. Formally, the derivative of the function f' at a is the limit

$$f'(a) = \lim_{h\to 0}\frac{f(a+h)-f(a)}{h}$$

of the difference quotient as h approaches zero, if this limit exists. If the limit exists, then f is *differentiable* at a. Here f (a) is one of several common notations for the derivative.

Equivalently, the derivative satisfies the property that

$$\lim_{h\to 0}\frac{f(a+h)-f(a)-f'(a)\cdot h}{h} = 0,$$

which has the intuitive interpretation that the tangent line to f at a gives the *best linear approximation*

$$f(a+h) \approx f(a) + f'(a)h$$

to f near a (i.e., for small h). This interpretation is the easiest to generalize to other settings.

Substituting 0 for h in the difference quotient causes division by zero, so the slope of the tangent line cannot be found directly using this method. Instead, define $Q(h)$ to be the difference quotient as a function of h:

$$Q(h) = \frac{f(a+h)-f(a)}{h}.$$

$Q(h)$ is the slope of the secant line between $(a, f(a))$ and $(a + h, f(a + h))$. If f is a continuous function, meaning that its graph is an unbroken curve with no gaps, then Q is a continuous function away from $h = 0$. If the limit $\lim_{h\to 0} Q(h)$ exists, meaning that there is a way of choosing a value for $Q(0)$ that makes the graph of Q a continuous function, then the function f is differentiable at a, and its derivative at a equals $Q(0)$.

In practice, the existence of a continuous extension of the difference quotient $Q(h)$ to $h = 0$ is shown by modifying the numerator to cancel h in the denominator. Such manipulations can make the limiting value of Q for small h clear even though Q is still not defined at $h = 0$. This process can be long and tedious for complicated functions, and many shortcuts are commonly used to simplify the process.

Example

The squaring function $f(x) = x^2$ is differentiable at $x = 3$, and its derivative there is 6. This result is established by calculating the limit as h approaches zero of the difference quotient of $f(3)$:

$$f'(3) = \lim_{h\to 0}\frac{f(3+h)-f(3)}{h} = \lim_{h\to 0}\frac{(3+h)^2-3^2}{h} = \lim_{h\to 0}\frac{9+6h+h^2-9}{h} = \lim_{h\to 0}\frac{6h+h^2}{h} = \lim_{h\to 0}(6+h).$$

The last expression shows that the difference quotient equals $6 + h$ when $h \neq 0$ and is undefined when $h = 0$, because of the definition of the difference quotient. However, the definition of the limit says the difference quotient does not need to be defined when $h = 0$. The limit is the result of letting h go to zero, meaning it is the value that $6 + h$ tends to as h becomes very small:

$$\lim_{h\to 0}(6+h) = 6+0 = 6.$$

Hence the slope of the graph of the squaring function at the point (3, 9) is 6, and so its derivative at $x = 3$ is $f'(3) = 6$.

More generally, a similar computation shows that the derivative of the squaring function at $x = a$ is $f'(a) = 2a$.

Continuity and Differentiability

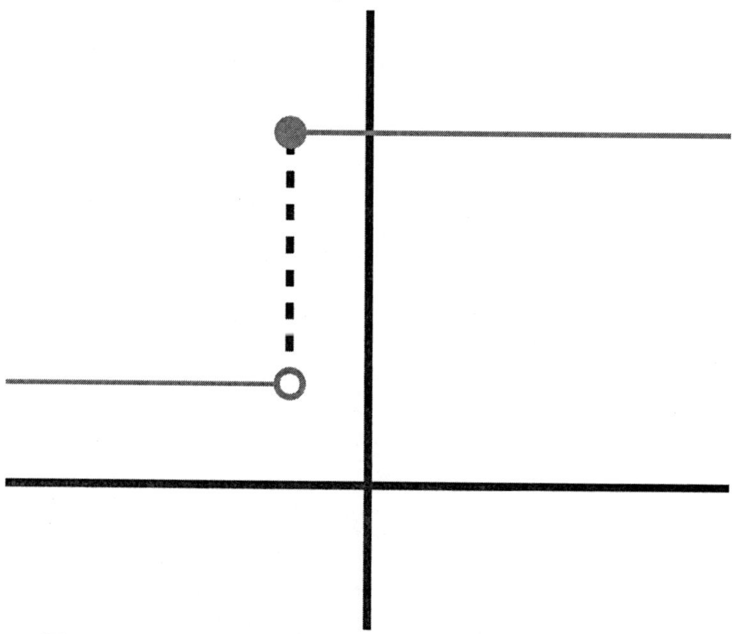

Figure: This function does not have a derivative at the marked point, as the function is not continuous there.

If $y = f(x)$ is differentiable at a, then f must also be continuous at a. As an example, choose a point a and let f be the step function that returns a value, say 1, for all x less than a, and returns a different value, say 10, for all x greater than or equal to a. f cannot have a derivative at a. If h is negative, then $a + h$ is on the low part of the step, so the secant line from a to $a + h$ is very steep, and as h tends to zero the slope tends to infinity. If h is positive, then $a + h$ is on the high part of the step, so the secant line from a to $a + h$ has slope zero. Consequently the secant lines do not approach any single slope, so the limit of the difference quotient does not exist.

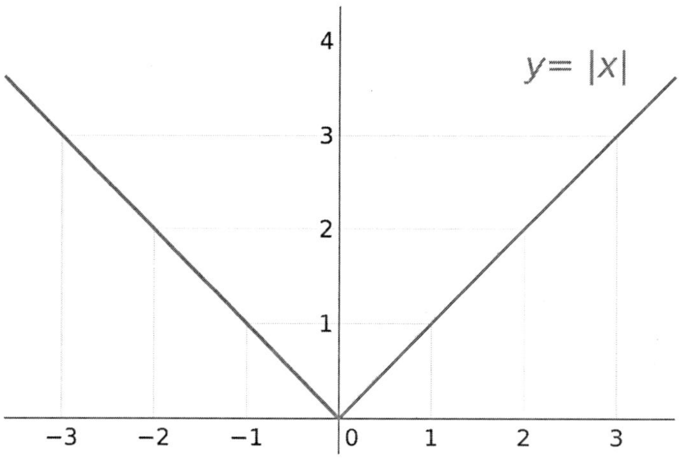

Figure: *The absolute value function is continuous, but fails to be differentiable at $x = 0$ since the tangent slopes do not approach the same value from the left as they do from the right.*

However, even if a function is continuous at a point, it may not be differentiable there. For example, the absolute value function $y = |x|$ is continuous at $x = 0$, but it is not differentiable there. If h is positive, then the slope of the secant line from 0 to h is one, whereas if h is negative, then the slope of the secant line from 0 to h is negative one. This can be seen graphically as a "kink" or a "cusp" in the graph at $x = 0$. Even a function with a smooth graph is not differentiable at a point where its tangent is vertical: For instance, the function $y = x^{1/3}$ is not differentiable at $x = 0$.

In Summary: for a function f to have a derivative it is *necessary* for the function f to be continuous, but continuity alone is not *sufficient*.

Most functions that occur in practice have derivatives at all points or at almost every point. Early in the history of calculus, many mathematicians assumed that a continuous function was differentiable at most points. Under mild conditions, for example if the function is a monotone function or a Lipschitz function, this is true. However, in 1872 Weierstrass found the first example of a function that is continuous everywhere but differentiable nowhere. This example is now known as the Weierstrass function. In 1931, Stefan Banach proved that the set of functions that have a derivative at some point is a meager set in the space of all continuous functions. Informally, this means that hardly any continuous functions have a derivative at even one point.

The Derivative as a Function

Let f be a function that has a derivative at every point a in the domain of f. Because every point a has a derivative, there is a function that sends the point a to the derivative of f at a. This function is written $f2\ (x)$ and is called the *derivative function* or the *derivative* of f. The derivative of f collects all the derivatives of f at all the points in the domain of f. Sometimes f has a derivative at most, but not all, points of its domain. The function whose value at a equals $f'(a)$ whenever $f'(a)$ is defined and elsewhere is undefined is also called the derivative of f. It is still a function, but its domain is strictly smaller than the domain of f.

Using this idea, differentiation becomes a function of functions: The derivative is an operator whose domain is the set of all functions that have derivatives at every point of their domain and whose range is a set of functions. If we denote this operator by D, then $D(f)$ is the function $f'(x)$. Since $D(f)$ is a function, it can be evaluated at a point a. By the definition of the derivative function, $D(f)(a) = f'(a)$. For comparison, consider the doubling function $f(x) = 2x$; f is a real-valued function of a real number, meaning that it takes numbers as inputs and has numbers as outputs:

$$\begin{aligned} 1 &\mapsto 2, \\ 2 &\mapsto 4, \\ 3 &\mapsto 6. \end{aligned}$$

The operator D, however, is not defined on individual numbers. It is only defined on functions:

$$\begin{aligned} D(x \mapsto 1) &= (x \mapsto 0), \\ D(x \mapsto x) &= (x \mapsto 1), \\ D(x \mapsto x^2) &= (x \mapsto 2 \cdot x). \end{aligned}$$

Because the output of D is a function, the output of D can be evaluated at a point. For instance, when D is applied to the squaring function,

$$x \mapsto x^2,$$

D outputs the doubling function,

$$x \mapsto 2x,$$

which we named $f(x)$. This output function can then be evaluated to get $f(1) = 2$, $f(2) = 4$, and so on.

Higher Derivatives

Let f be a differentiable function, and let $f'(x)$ be its derivative. The derivative of $f'(x)$ (if it has one) is written $f''(x)$ and is called the *second derivative of f*. Similarly, the derivative of a second derivative, if it exists, is written $f'''(x)$ and is called the *third derivative of f*. These repeated derivatives are called *higher-order derivatives.*

If $x(t)$ represents the position of an object at time t, then the higher-order derivatives of x have physical interpretations. The second derivative of x is the derivative of $x2$ (t), the velocity, and by definition this is the object's acceleration. The third derivative of x is defined to be the jerk, and the fourth derivative is defined to be the jounce.

A function f need not have a derivative, for example, if it is not continuous. Similarly, even if f does have a derivative, it may not have a second derivative. For example, let

$$f(x) = \begin{cases} +x^2, & \text{if } x \geq 0 \\ -x^2, & \text{if } x \leq 0. \end{cases}$$

Calculation shows that f is a differentiable function whose derivative is

$$f'(x) = \begin{cases} +2x, & \text{if } x \geq 0 \\ -2x, & \text{if } x \leq 0. \end{cases}$$

$f'(x)$ is twice the absolute value function, and it does not have a derivative at zero. Similar examples show that a function can have k derivatives for any non-negative integer k but no $(k + 1)$-order derivative. A function that has k successive derivatives is called *k times differentiable*. If in addition the kth derivative is continuous, then the function is said to be of differentiability class C^k. (This is a stronger condition than having k derivatives.) A function that has infinitely many derivatives is called *infinitely differentiable* or *smooth.*

On the real line, every polynomial function is infinitely differentiable. By standard differentiation rules, if a polynomial of degree n is differentiated n times, then it becomes a constant function. All of its subsequent derivatives are identically zero. In particular, they exist, so polynomials are smooth functions.

The derivatives of a function f at a point x provide polynomial approximations to that function near x. For example, if f is twice differentiable, then

$$f(x+h) \approx f(x) + f'(x)h + \tfrac{1}{2} f''(x)h^2$$

in the sense that

$$\lim_{h \to 0} \frac{f(x+h) - f(x) - f'(x)h - \dfrac{1}{2} f''(x)h^2}{h^2} = 0.$$

If f is infinitely differentiable, then this is the beginning of the Taylor series for f evaluated at $x+h$ around x.

Inflection Point

A point where the second derivative of a function changes sign is called an *inflection point*. At an inflection point, the second derivative may be zero, as in the case of the inflection point $x=0$ of the function $y=x^3$, or it may fail to exist, as in the case of the inflection point $x=0$ of the function $y=x^{1/3}$. At an inflection point, a function switches from being a convex function to being a concave function or vice versa.

Notations for Differentiation

Leibniz's Notation: The notation for derivatives introduced by Gottfried Leibniz is one of the earliest. It is still commonly used when the equation $y = f(x)$ is viewed as a functional relationship between dependent and independent variables. Then the first derivative is denoted by

$$\frac{dy}{dx}, \quad \frac{df}{dx}(x), \text{ or } \frac{d}{dx} f(x),$$

and was once thought of as an infinitesimal quotient. Higher derivatives are expressed using the notation

$$\frac{d^n y}{dx^n}, \quad \frac{d^n f}{dx^n}(x), \text{ or } \frac{d^n}{dx^n} f(x)$$

for the nth derivative of $y = f(x)$ (with respect to x). These are abbreviations for multiple applications of the derivative operator. For example,

$$\frac{d^2y}{dx^2} = \frac{d}{dx}\left(\frac{dy}{dx}\right).$$

With Leibniz's notation, we can write the derivative of y at the point $x = a$ in two different ways:

$$\left.\frac{dy}{dx}\right|_{x=a} = \frac{dy}{dx}(a).$$

Leibniz's notation allows one to specify the variable for differentiation (in the denominator). This is especially relevant for partial differentiation. It also makes the chain rule easy to remember:

$$\frac{dy}{dx} = \frac{dy}{du} \cdot \frac{du}{dx}.$$

Lagrange's Notation

Sometimes referred to as *prime notation*, one of the most common modern notations for differentiation is due to Joseph-Louis Lagrange and uses the prime mark, so that the derivative of a function $f(x)$ is denoted $f'(x)$ or simply f'. Similarly, the second and third derivatives are denoted

$$(f')' = f'' \quad \text{and} \quad (f'')' = f'''.$$

To denote the number of derivatives beyond this point, some authors use Roman numerals in superscript, whereas others place the number in parentheses:

$$f^{\text{iv}} \quad \text{or} \quad f^{(4)}.$$

The latter notation generalizes to yield the notation $f^{(n)}$ for the nth derivative of f — this notation is most useful when we wish to talk about the derivative as being a function itself, as in this case the Leibniz notation can become cumbersome.

Newton's Notation

Newton's notation for differentiation, also called the dot notation, places a dot over the function name to represent a time derivative. If $y = f(t)$, then

$$\dot{y} \text{ and } \ddot{y}$$

denote, respectively, the first and second derivatives of y with respect to t. This notation is used exclusively for time derivatives, meaning that the independent variable of the function represents time. It is very common in physics and in mathematical disciplines connected with physics such as differential equations. While the notation becomes unmanageable for high-order derivatives, in practice only very few derivatives are needed.

Euler's Notation

Euler's notation uses a differential operator D, which is applied to a function f to give the first derivative Df. The second derivative is denoted D^2f, and the nth derivative is denoted D^nf.

If $y = f(x)$ is a dependent variable, then often the subscript x is attached to the D to clarify the independent variable x. Euler's notation is then written

$$D_x y \quad \text{or} \quad D_x f(x) \,,$$

although this subscript is often omitted when the variable x is understood, for instance when this is the only variable present in the expression. Euler's notation is useful for stating and solving linear differential equations.

Computing the Derivative

The derivative of a function can, in principle, be computed from the definition by considering the difference quotient, and computing its limit. In practice, once the derivatives of a few simple functions are known, the derivatives of other functions are more easily computed using *rules* for obtaining derivatives of more complicated functions from simpler ones.

Derivatives of Elementary Functions

Most derivative computations eventually require taking the derivative of some common functions. The following incomplete list gives some of the most frequently used functions of a single real variable and their derivatives.

- *Derivatives of powers*: if

  $$f(x) = x^r,$$

  where r is any real number, then

  $$f'(x) = rx^{r-1},$$

  wherever this function is defined. For example, if $f(x) = x^{1/4}$, then

  $$f'(x) = (1/4)x^{-3/4},$$

and the derivative function is defined only for positive x, not for $x = 0$. When $r = 0$, this rule implies that $f'x)$ is zero for $x \neq 0$, which is almost the constant rule (stated below).

- *Exponential and logarithmic functions*:

$$\frac{d}{dx}e^x = e^x.$$

$$\frac{d}{dx}a^x = \ln(a)a^x.$$

$$\frac{d}{dx}\ln(x) = \frac{1}{x}, \qquad x > 0.$$

$$\frac{d}{dx}\log_a(x) = \frac{1}{x\ln(a)}.$$

- *Trigonometric functions*:

$$\frac{d}{dx}\sin(x) = \cos(x).$$

$$\frac{d}{dx}\cos(x) = -\sin(x).$$

$$\frac{d}{dx}\tan(x) = \sec^2(x) = \frac{1}{\cos^2(x)} = 1 + \tan^2(x).$$

- *Inverse trigonometric functions*:

$$\frac{d}{dx}\arcsin(x) = \frac{1}{\sqrt{1-x^2}}, -1 < x < 1.$$

$$\frac{d}{dx}\arccos(x) = -\frac{1}{\sqrt{1-x^2}}, -1 < x < 1.$$

$$\frac{d}{dx}\arctan(x) = \frac{1}{1+x^2}$$

Rules for Finding the Derivative

In many cases, complicated limit calculations by direct application of Newton's difference quotient can be avoided using differentiation rules. Some of the most basic rules are the following.

- *Constant rule*: if $f(x)$ is constant, then $f' = 0$.
- *Sum rule*:

$(\alpha f + \beta g)' = \alpha f' + \beta g'$ for all functions f and g and all real numbers α and β.

- *Product rule*:

$(fg)' = f'g + fg'$ for all functions f and g. By extension, this means that the derivative of a constant times a function is the constant times the derivative of the function: $\frac{d}{dr}\pi r^2 = 2\pi r$.

- *Quotient rule*:

$\left(\frac{f}{g}\right)' = \frac{f'g - fg'}{g^2}$ for all functions f and g at all inputs where $g \neq 0$.

- *Chain rule*: If $f(x) = h(g(x))$, then

$$f'(x) = h'(g(x)) \cdot g'(x).$$

Example Computation

The derivative of

$$f(x) = x^4 + \sin(x^2) - \ln(x)e^x + 7$$

is

$$\begin{aligned} f'(x) &= 4x^{(4-1)} + \frac{d\left(x^2\right)}{dx}\cos(x^2) - \frac{d(\ln x)}{dx}e^x - \ln x\frac{d\left(e^x\right)}{dx} + 0 \\ &= 4x^3 + 2x\cos(x^2) - \frac{1}{x}e^x - \ln(x)e^x. \end{aligned}$$

Here the second term was computed using the chain rule and third using the product rule. The known derivatives of the elementary functions x^2, x^4, $\sin(x)$, $\ln(x)$ and $\exp(x) = e^x$, as well as the constant 7, were also used.

Derivatives in Higher Dimensions

Derivatives of Vector Valued Functions: A vector-valued function $y(t)$ of a real variable sends real numbers to vectors in some vector space R^n.

A vector-valued function can be split up into its coordinate functions $y_1(t)$, $y_2(t)$, ..., $y_n(t)$, meaning that $y(t) = (y_1(t), ..., y_n(t))$. This includes, for example, parametric curves in R^2 or R^3. The coordinate functions are real valued functions, so the above definition of derivative applies to them. The derivative of $y(t)$ is defined to be the vector, called the tangent vector, whose coordinates are the derivatives of the coordinate functions. That is,

$$y'(t) = (y_1'(t), \ldots, y_n'(t)).$$

Equivalently,

$$y'(t) = \lim_{h \to 0} \frac{y(t+h) - y(t)}{h},$$

if the limit exists. The subtraction in the numerator is subtraction of vectors, not scalars. If the derivative of y exists for every value of t, then y2 is another vector valued function.

If e_1, ..., e_n is the standard basis for R^n, then y(t) can also be written as $y_1(t)e_1 + ... + y_n(t)e_n$. If we assume that the derivative of a vector-valued function retains the linearity property, then the derivative of y(t) must be

$$y_1'(t)e_1 + \cdots + y_n'(t)e_n$$

because each of the basis vectors is a constant.

This generalization is useful, for example, if y(t) is the position vector of a particle at time t; then the derivative y′ (t) is the velocity vector of the particle at time t.

Partial Derivatives

Suppose that f is a function that depends on more than one variable. For instance,

$$f(x, y) = x^2 + xy + y^2.$$

f can be reinterpreted as a family of functions of one variable indexed by the other variables:

$$f(x, y) = f_x(y) = x^2 + xy + y^2.$$

In other words, every value of x chooses a function, denoted f_x, which is a function of one real number. That is,

$$x \mapsto f_x,$$

$$f_x(y) = x^2 + xy + y^2.$$

Once a value of x is chosen, say a, then $f(x,y)$ determines a function f_a that sends y to a^2 + ay + y^2:

$$f_a(y) = a^2 + ay + y^2.$$

In this expression, a is a *constant*, not a *variable*, so f_a is a function of only one real variable. Consequently the definition of the derivative for a function of one variable applies:

$$f_{a'}(y) = a + 2y.$$

The above procedure can be performed for any choice of a. Assembling the derivatives together into a function gives a function that describes the variation of f in the y direction:

$$\frac{\partial f}{\partial y}(x,y) = x + 2y.$$

This is the partial derivative of f with respect to y. Here ∂ is a rounded d called the partial derivative symbol. To distinguish it from the letter d, " is sometimes pronounced "der", "del", or "partial" instead of "dee".

In general, the partial derivative of a function $f(x_1, \ldots, x_n)$ in the direction x_i at the point $(a_1 \ldots, a_n)$ is defined to be:

$$\frac{\partial f}{\partial x_i}(a_1,\ldots,a_n) = \lim_{h\to 0}\frac{f(a_1,\ldots,a_i+h,\ldots,a_n) - f(a_1,\ldots,a_i,\ldots,a_n)}{h}.$$

In the above difference quotient, all the variables except x_i are held fixed. That choice of fixed values determines a function of one variable

$$f_{a_1,\ldots,a_{i-1},a_{i+1},\ldots,a_n}(x_i) = f(a_1,\ldots,a_{i-1},x_i,a_{i+1},\ldots,a_n),$$

and, by definition,

$$\frac{df_{a_1,\ldots,a_{i-1},a_{i+1},\ldots,a_n}}{dx_i}(a_i) = \frac{\partial f}{\partial x_i}(a_1,\ldots,a_n).$$

In other words, the different choices of a index a family of one-variable functions just as in the example above. This expression also shows that the computation of partial derivatives reduces to the computation of one-variable derivatives.

An important example of a function of several variables is the case of a scalar-valued function $f(x_1,\ldots x_n)$ on a domain in Euclidean space R^n (e.g., on R^2 or R^3). In this case f has a partial derivative $\partial f/\partial x_j$ with respect to each variable x_j. At the point a, these partial derivatives define the vector

$$\nabla f(a) = \left(\frac{\partial f}{\partial x_1}(a),\ldots,\frac{\partial f}{\partial x_n}(a)\right).$$

This vector is called the gradient of f at a. If f is differentiable at every point in some domain, then the gradient is a vector-valued function ∇f that takes the point a to the vector $\nabla f(a)$. Consequently the gradient determines a vector field.

Directional Derivatives

If f is a real-valued function on R^n, then the partial derivatives of f measure its variation in the direction of the coordinate axes. For example, if f is a function of x and y, then its partial derivatives measure the variation in f in the x direction and the y direction. They do not, however, directly measure the variation of f in any other direction, such as along the diagonal line $y = x$. These are measured using directional derivatives. Choose a vector

$$\mathrm{v} = (v_1, \ldots, v_n).$$

The directional derivative of f in the direction of v at the point x is the limit

$$D_{\mathrm{v}} f(\mathrm{x}) = \lim_{h \to 0} \frac{f(\mathrm{x} + h\mathrm{v}) - f(\mathrm{x})}{h}.$$

In some cases it may be easier to compute or estimate the directional derivative after changing the length of the vector. Often this is done to turn the problem into the computation of a directional derivative in the direction of a unit vector. To see how this works, suppose that $\mathrm{v} = \lambda\mathrm{u}$. Substitute $h = k/\lambda$ into the difference quotient. The difference quotient becomes:

$$\frac{f(\mathrm{x} + (k/\lambda)(\lambda\mathrm{u})) - f(\mathrm{x})}{k/\lambda} = \lambda \cdot \frac{f(\mathrm{x} + k\mathrm{u}) - f(\mathrm{x})}{k}.$$

This is λ times the difference quotient for the directional derivative of f with respect to u. Furthermore, taking the limit as h tends to zero is the same as taking the limit as k tends to zero because h and k are multiples of each other. Therefore $D_{\mathrm{v}}(f) = \lambda D_{\mathrm{u}}(f)$. Because of this rescaling property, directional derivatives are frequently considered only for unit vectors. If all the partial derivatives of f exist and are continuous at x, then they determine the directional derivative of f in the direction v by the formula:

$$D_{\mathrm{v}} f(\mathrm{x}) = \sum_{j=1}^{n} v_j \frac{\partial f}{\partial x_j}.$$

This is a consequence of the definition of the total derivative. It follows that the directional derivative is linear in v, meaning that $D_{\mathrm{v}+\mathrm{w}}(f) = D_{\mathrm{v}}(f) + D_{\mathrm{w}}(f)$.

The same definition also works when f is a function with values in R^m. The above definition is applied to each component of the vectors. In this case, the directional derivative is a vector in R^m.

Total Derivative, Total Differential and Jacobian Matrix

When f is a function from an open subset of R^n to R^m, then the directional derivative of f in a chosen direction is the best linear approximation to f at that point and in that direction. But when $n > 1$, no single directional derivative can give a complete picture of the behaviour of f. The total derivative, also called the (total) differential, gives a complete picture by considering all directions at once. That is, for any vector v starting at a, the linear approximation formula holds:

$$f(a+v) \approx f(a) + f'(a)v.$$

Just like the single-variable derivative, $f'(a)$ is chosen so that the error in this approximation is as small as possible.

If n and m are both one, then the derivative $f'(a)$ is a number and the expression $f'(a)v$ is the product of two numbers. But in higher dimensions, it is impossible for $f'(a)$ to be a number. If it were a number, then $f'(a)v$ would be a vector in R^n while the other terms would be vectors in R^m, and therefore the formula would not make sense. For the linear approximation formula to make sense, $f'(a)$ must be a function that sends vectors in R^n to vectors in R^m, and $f'(a)v$ must denote this function evaluated at v.

To determine what kind of function it is, notice that the linear approximation formula can be rewritten as

$$f(a+v) - f(a) \approx f'(a)v.$$

Notice that if we choose another vector w, then this approximate equation determines another approximate equation by substituting w for v. It determines a third approximate equation by substituting both w for v and a + v for a. By subtracting these two new equations, we get

$$f(a+v+w) - f(a+v) - f(a+w) + f(a) \approx f'(a+v)w - f'(a)w.$$

If we assume that v is small and that the derivative varies continuously in a, then $f'(a + v)$ is approximately equal to $f'(a)$, and therefore the right-hand side is approximately zero. The left-hand side can be rewritten in a different way using the linear approximation formula with v + w substituted for v. The linear approximation formula implies:

$$\begin{aligned} 0 &\approx f(a+v+w) - f(a+v) - f(a+w) + f(a) \\ &= (f(a+v+w) - f(a)) - (f(a+v) - f(a)) - (f(a+w) - f(a)) \\ &\approx f'(a)(v+w) - f'(a)v - f'(a)w. \end{aligned}$$

This suggests that f' (a) is a linear transformation from the vector space R^n to the vector space R^m. In fact, it is possible to make this a precise derivation by measuring the error in the approximations. Assume that the error in these linear approximation formula is bounded by a constant times $\|v\|$, where the constant is independent of v but depends continuously on a. Then, after adding an appropriate error term, all of the above approximate equalities can be rephrased as inequalities. In particular, f' (a) is a linear transformation up to a small error term. In the limit as v and w tend to zero, it must therefore be a linear transformation. Since we define the total derivative by taking a limit as v goes to zero, f' (a) must be a linear transformation.

In one variable, the fact that the derivative is the best linear approximation is expressed by the fact that it is the limit of difference quotients. However, the usual difference quotient does not make sense in higher dimensions because it is not usually possible to divide vectors. In particular, the numerator and denominator of the difference quotient are not even in the same vector space: The numerator lies in the codomain R^m while the denominator lies in the domain R^n. Furthermore, the derivative is a linear transformation, a different type of object from both the numerator and denominator. To make precise the idea that f' (a) is the best linear approximation, it is necessary to adapt a different formula for the one-variable derivative in which these problems disappear. If f: R → R, then the usual definition of the derivative may be manipulated to show that the derivative of f at a is the unique number f' (a) such that

$$\lim_{h\to 0}\frac{f(a+h)-f(a)-f'(a)h}{h}=0.$$

This is equivalent to

$$\lim_{h\to 0}\frac{|f(a+h)-f(a)-f'(a)h|}{|h|}=0$$

because the limit of a function tends to zero if and only if the limit of the absolute value of the function tends to zero. This last formula can be adapted to the many-variable situation by replacing the absolute values with norms. The definition of the total derivative of f at a, therefore, is that it is the unique linear transformation f' (a) : $R^n \to R^m$ such that

$$\lim_{\mathrm{h}\to 0}\frac{\|f(\mathrm{a}+\mathrm{h})-f(\mathrm{a})-f'(\mathrm{a})\mathrm{h}\|}{\|\mathrm{h}\|}=0.$$

Here h is a vector in R^n, so the norm in the denominator is the standard length on R^n. However, f' (a)h is a vector in R^m, and the norm in the numerator is the standard length on R^m. If v is a vector starting at a, then f' (a)v is called the pushforward of v by f and is sometimes written f_*v.

If the total derivative exists at a, then all the partial derivatives and directional derivatives of f exist at a, and for all v, f' (a)v is the directional derivative of f in the direction v. If we write f using coordinate functions, so that $f = (f_1, f_2, ..., f_m)$, then the total derivative can be expressed using the partial derivatives as a matrix. This matrix is called the Jacobian matrix of f at a:

$$f'(\mathrm{a}) = \mathrm{Jac}_{\mathrm{a}} = \left(\frac{\partial f_i}{\partial x_j}\right)_{ij}.$$

The existence of the total derivative f' (a) is strictly stronger than the existence of all the partial derivatives, but if the partial derivatives exist and are continuous, then the total derivative exists, is given by the Jacobian, and depends continuously on a.

The definition of the total derivative subsumes the definition of the derivative in one variable. That is, if f is a real-valued function of a real variable, then the total derivative exists if and only if the usual derivative exists. The Jacobian matrix reduces to a 1×1 matrix whose only entry is the derivative f' (x). This 1×1 matrix satisfies the property that $f(a + h) - f(a) - f'(a)h$ is approximately zero, in other words that

$$f(a+h) \approx f(a) + f'(a)h.$$

Up to changing variables, this is the statement that the function $x \mapsto f(a) + f'(a)(x-a)$ is the best linear approximation to f at a.

The total derivative of a function does not give another function in the same way as the one-variable case. This is because the total derivative of a multivariable function has to record much more information than the derivative of a single-variable function. Instead, the total derivative gives a function from the tangent bundle of the source to the tangent bundle of the target.

The natural analog of second, third, and higher-order total derivatives is not a linear transformation, is not a function on the tangent bundle, and is not built by repeatedly taking the total derivative. The analog of a higher-order derivative, called a jet, cannot be a linear transformation because higher-order derivatives reflect subtle geometric information, such as concavity, which cannot be described in terms of linear data such as vectors. It cannot be a function on the tangent bundle because the tangent bundle only has room for the base space and the directional derivatives. Because jets capture higher-order information, they take as arguments additional coordinates representing higher-order changes in direction. The space determined by these additional coordinates is called the jet bundle. The relation between the total derivative and the partial derivatives of a function is paralleled in the relation between the kth order jet of a function and its partial derivatives of order less than or equal to k.

By repeatedly taking the total derivative, one obtains higher versions of the Fréchet derivative, specialized to R^p. The kth order total derivative may be interpreted as a map

$$D^k f : \mathbb{R}^n \to L^k(\mathbb{R}^n \times \cdots \times \mathbb{R}^n, \mathbb{R}^m)$$

which takes a point x in R^n and assigns to it an element of the space of k-linear maps from R^n to R^m- the "best" (in a certain precise sense) k-linear approximation to f at that point. By precomposing it with the diagonal map Δ, x→(x, x), a generalized Taylor series may be begun as

$$\begin{aligned} f(\mathrm{x}) &\approx f(\mathrm{a}) + (Df)(\mathrm{x}) + (D^2 f)(\Delta(\mathrm{x}-\mathrm{a})) + \cdots \\ &= f(\mathrm{a}) + (Df)(\mathrm{x}-\mathrm{a}) + (D^2 f)(\mathrm{x}-\mathrm{a}, \mathrm{x}-\mathrm{a}) + \cdots \\ &= f(\mathrm{a}) + \sum_i (Df)_i (\mathrm{x}-\mathrm{a})^i + \sum_{j,k} (D^2 f)_{jk} (\mathrm{x}-\mathrm{a})^j (\mathrm{x}-\mathrm{a})^k + \cdots \end{aligned}$$

where f(a) is identified with a constant function, $(x\text{-}a)^i$ are the components of the vector x-a, and $(D f)_i$ and $(D^2 f)_{jk}$ are the components of $D f$ and $D^2 f$ as linear transformations.

Generalizations

The concept of a derivative can be extended to many other settings. The common thread is that the derivative of a function at a point serves as a linear approximation of the function at that point.

- An important generalization of the derivative concerns complex functions of complex variables, such as functions from (a domain in) the complex numbers C to C. The notion of the derivative of such a function is obtained by replacing real variables with complex variables in the definition. If C is identified with R^2

by writing a complex number z as $x + i y$, then a differentiable function from C to C is certainly differentiable as a function from R^2 to R^2 (in the sense that its partial derivatives all exist), but the converse is not true in general: the complex derivative only exists if the real derivative is *complex linear* and this imposes relations between the partial derivatives called the Cauchy Riemann equations.

- Another generalization concerns functions between differentiable or smooth manifolds. Intuitively speaking such a manifold M is a space that can be approximated near each point x by a vector space called its tangent space: the prototypical example is a smooth surface in R^3. The derivative (or differential) of a (differentiable) map $f: M \to N$ between manifolds, at a point x in M, is then a linear map from the tangent space of M at x to the tangent space of N at $f(x)$. The derivative function becomes a map between the tangent bundles of M and N. This definition is fundamental in differential geometry and has many uses (differential) and pullback (differential geometry).
- Differentiation can also be defined for maps between infinite dimensional vector spaces such as Banach spaces and Fréchet spaces. There is a generalization both of the directional derivative, called the Gâteaux derivative, and of the differential, called the Fréchet derivative.
- One deficiency of the classical derivative is that not very many functions are differentiable. Nevertheless, there is a way of extending the notion of the derivative so that all continuous functions and many other functions can be differentiated using a concept known as the weak derivative. The idea is to embed the continuous functions in a larger space called the space of distributions and only require that a function is differentiable "on average".
- The properties of the derivative have inspired the introduction and study of many similar objects in algebra and topology.
- The discrete equivalent of differentiation is finite differences. The study of differential calculus is unified with the calculus of finite differences in time scale calculus.

Transcendental Function

A transcendental function is a function that is not algebraic. Such a function cannot be expressed as a solution of a polynomial equation whose coefficients are themselves polynomials with rational coefficients. Examples of transcendental functions include the exponential function, the logarithm, and the trigonometric functions.

Definition

Formally, an analytic function $f(z)$ of the real or complex variables $z_1,\ldots,z_n$ is transcendental if $z_1, \ldots, z_n, f(z)$ are algebraically independent, i.e., if f is transcendental over the field $C(z_1, \ldots, z_n)$.

A transcendental function is a function that "transcends" algebra in the sense that it cannot be expressed in terms of a finite sequence of the algebraic operations of addition, multiplication, power, and root extraction.

Examples

$$f_1(x) = x^{\pi}$$

$$f_2(x) = c^x, c \neq 0,1$$

$$f_3(x) = x^x = {}^2x$$

$$f_4(x) = x^{\frac{1}{x}}$$

$$f_5(x) = \log_c x, c \neq 0,1$$

$$f_6(x) = \sin x$$

Note that in particular for f_2 if we set c equal to e, the base of the natural logarithm, then we get that e^x is a transcendental function. Similarly, if we set c equal to e in f_5, then we get that $\ln(x)$, the natural logarithm, is a transcendental function.

Algebraic and Transcendental Functions

The logarithm and the exponential function are examples of transcendental functions. *Transcendental function* is a term often used to describe the trigonometric functions (sine, cosine, tangent, their reciprocals cotangent, secant, and cosecant, the now little-used versine, haversine, and coversine, their analogs the hyperbolic functions and so forth).

A function that is not transcendental is said to be algebraic. Examples of algebraic functions are rational functions and the square root function. The operation of taking the indefinite integral of an algebraic function is a source of transcendental functions. For example, the logarithm function arose from the reciprocal function in an effort to find the area of a hyperbolic sector. Thus the hyperbolic angle and the hyperbolic functions sinh, cosh, and tanh are all transcendental. Differential algebra examines how integration frequently creates functions that are algebraically independent of some class, such as when one takes polynomials with trigonometric functions as variables.

Dimensional Analysis

In dimensional analysis, transcendental functions are notable because they make sense only when their argument is dimensionless (possibly after algebraic reduction). Because of this, transcendental functions can be an easy-to-spot source of dimensional errors. For example, log(5 metres) is a nonsensical expression, unlike log(5 metres / 3 metres) or log(3) metres. One could attempt to apply a logarithmic identity to get log(10) + log(m), which highlights the problem: applying a non-algebraic operation to a dimension creates meaningless results.

Exceptional Set

If $f(z)$ is an algebraic function and α is an algebraic number then $f(\alpha)$ will also be an algebraic number. The converse is not true: there are entire transcendental functions $f(z)$ such that $f(\alpha)$ is an algebraic number for any algebraic α.

In many instances, however, the set of algebraic numbers α where $f(\alpha)$ is algebraic is fairly small. For example, if f is the exponential function, $f(z) = e^z$, then the only algebraic number α where $f(\alpha)$ is also algebraic is $\alpha = 0$, where $f(\alpha) = 1$. For a given transcendental function this set of algebraic numbers giving algebraic results is called the exceptional set of the function, that is the set

$$\mathcal{E}(f) = \{\alpha \in \overline{\mathbb{Q}} : f(\alpha) \in \overline{\mathbb{Q}}\}.$$

If this set can be calculated then it can often lead to results in transcendence theory. For example, Lindemann proved in 1882 that the exceptional set of the exponential function is just {0}. In particular exp(1) = e is transcendental. Also, since exp($i\pi$) = -1 is algebraic we know that $i\pi$ cannot be algebraic. Since i is algebraic this implies that π is a transcendental number.

In general, finding the exceptional set of a function is a difficult problem, but it has been calculated for some functions:

- $\mathcal{E}(\exp) = \{0\}$,
- $\mathcal{E}(j) = \{\alpha \in \mathrm{H} : [\mathrm{Q}(\alpha) : \mathrm{Q}] = 2\}$,

- o Here j is Klein's j-invariant, H is the upper half-plane, and [Q(α): Q] is the degree of the number field Q(α). This result is due to Theodor Schneider.
- $\mathcal{E}(2^x) = \mathrm{Q}$,
    - o This result is a corollary of the Gelfond–Schneider theorem which says that if a is algebraic and not 0 or 1, and if β is algebraic and irrational then a^β is transcendental. Thus the function 2^x could be replaced by c^x for any algebraic c not equal to 0 or 1. Indeed, we have:
- $\mathcal{E}(x^x) = \mathcal{E}(x^{\frac{1}{x}}) = \mathrm{Q} \setminus \{0\}$.
- A consequence of Schanuel's conjecture in transcendental number theory would be that $\mathcal{E}(e^{e^x}) = \varnothing$.
- A function with empty exceptional set that doesn't require one to assume this conjecture is the function $f(x) = \exp(1 + \pi x)$.

While calculating the exceptional set for a given function is not easy, it is known that given *any* subset of the algebraic numbers, say A, there is a transcendental function f whose exceptional set is A.[6] Since, as mentioned above, this includes taking A to be the whole set of algebraic numbers, there is no way to determine if a function is transcendental just by looking at its values at algebraic numbers. In fact, Alex Wilkie showed that the situation is even worse: he constructed a transcendental function f: R → R that is analytic everywhere but whose transcendence cannot be detected by any first-order method.

Total Derivative

In the mathematical field of differential calculus, the term total derivative has a number of closely related meanings.

- The total derivative (full derivative) of a function f, of several variables, e.g., t, x, y, etc., with respect to one of its input variables, e.g., t, is different from its partial derivative (). Calculation of the total derivative of f with respect to t does not assume that the other arguments are constant while t varies; instead, it allows the other arguments to depend on t. The total derivative adds in these *indirect dependencies* to find the overall dependency of f on t. For example, the total derivative of $f(t,x,y)$ with respect to t is

$$\frac{\mathrm{d}f}{\mathrm{d}t} = \frac{\partial f}{\partial t}\frac{\mathrm{d}t}{\mathrm{d}t} + \frac{\partial f}{\partial x}\frac{\mathrm{d}x}{\mathrm{d}t} + \frac{\partial f}{\partial y}\frac{\mathrm{d}y}{\mathrm{d}t}.$$

Which simplifies to

$$\frac{\mathrm{d}f}{\mathrm{d}t}=\frac{\partial f}{\partial t}+\frac{\partial f}{\partial x}\frac{\mathrm{d}x}{\mathrm{d}t}+\frac{\partial f}{\partial y}\frac{\mathrm{d}y}{\mathrm{d}t}.$$

Consider multiplying both sides of the equation by the differential $\mathrm{d}t$:

$$\mathrm{d}f=\frac{\partial f}{\partial t}\mathrm{d}t+\frac{\partial f}{\partial x}\mathrm{d}x+\frac{\partial f}{\partial y}\mathrm{d}y.$$

The result will be the differential change $\mathrm{d}f$ in the function *f*. Because *f* depends on *t*, some of that change will be due to the partial derivative of *t* with respect to . However, some of that change will also be due to the partial derivatives of *t* with respect to the variables *x* and *y*. So, the differential *dt* is applied to the total derivatives of *x* and *y* to find differentials d*x* and d*y*, which can then be used to find the contribution to d*f*.

- It refers to a differential operator such as

$$\frac{\mathrm{d}}{\mathrm{d}x}=\frac{\partial}{\partial x}+\sum_{j=1}^{k}\frac{\mathrm{d}y_j}{\mathrm{d}x}\frac{\partial}{\partial y_j},$$

which computes the total derivative of a function (with respect to *x* in this case).

- It refers to the (total) differential d*f* of a function, either in the traditional language of infinitesimals or the modern language of differential forms.
- A differential of the form

$$\sum_{j=1}^{k}f_j(x_1,\ldots,x_k)\,\mathrm{d}x_j$$

is called a total differential or an exact differential if it is the differential of a function. Again this can be interpreted infinitesimally, or by using differential forms and the exterior derivative.

- It is another name for the derivative as a linear map, i.e., if *f* is a differentiable function from R^n to R^m, then the (total) derivative (or differential) of *f* at $x\in\mathrm{R}^n$ is the linear map from R^n to R^m whose matrix is the Jacobian matrix of *f* at *x*.
- It is a synonym for the gradient, which is essentially the derivative of a function from R^n to R.
- It is sometimes used as a synonym for the material derivative, $\frac{Du}{Dt}$, in fluid mechanics.

Differentiation with Indirect Dependencies

Suppose that f is a function of two variables, x and y. Normally these variables are assumed to be independent. However, in some situations they may be dependent on each other. For example y could be a function of x, constraining the domain of f to a curve in R^2. In this case the partial derivative of f with respect to x does not give the true rate of change of f with respect to changing x because changing x necessarily changes y. The total derivative takes such dependencies into account.

For example, suppose

$$f(x,y) = xy .$$

The rate of change of f with respect to x is usually the partial derivative of f with respect to x; in this case,

$$\frac{\partial f}{\partial x} = y .$$

However, if y depends on x, the partial derivative does not give the true rate of change of f as x changes because it holds y fixed.

Suppose we are constrained to the line

$$y = x;$$

then

$$f(x,y) = f(x,x) = x^2 .$$

In that case, the total derivative of f with respect to x is

$$\frac{\mathrm{d}f}{\mathrm{d}x} = 2x .$$

Instead of immediately substituting for y in terms of x, this can be found equivalently using the chain rule:

$$\frac{\mathrm{d}f}{\mathrm{d}x} = \frac{\partial f}{\partial x} + \frac{\partial f}{\partial y}\frac{\mathrm{d}y}{\mathrm{d}x} = y + x \cdot 1 = x + x = 2x.$$

Notice that this is not equal to the partial derivative:

$$\frac{\mathrm{d}f}{\mathrm{d}x} = 2x \neq \frac{\partial f}{\partial x} = y = x .$$

While one can often perform substitutions to eliminate indirect dependencies, the chain rule provides for a more efficient and general technique. Suppose $M(t, p_1, ..., p_n)$ is a function of time t and n variables p_i which themselves depend on time. Then, the total time derivative of M is

$$\frac{\mathrm{d}M}{\mathrm{d}t}=\frac{\mathrm{d}}{\mathrm{d}t}M\big(t,p_1(t),\ldots,p_n(t)\big).$$

The chain rule for differentiating a function of several variables implies that

$$\frac{\mathrm{d}M}{\mathrm{d}t}=\frac{\partial M}{\partial t}+\sum_{i=1}^{n}\frac{\partial M}{\partial p_i}\frac{\mathrm{d}p_i}{\mathrm{d}t}=\Big(\frac{\partial}{\partial t}+\sum_{i=1}^{n}\frac{\mathrm{d}p_i}{\mathrm{d}t}\frac{\partial}{\partial p_i}\Big)(M).$$

This expression is often used in physics for a gauge transformation of the Lagrangian, as two Lagrangians that differ only by the total time derivative of a function of time and the n generalized coordinates lead to the same equations of motion. An interesting example concerns the resolution of causality concerning the Wheeler-Feynman time-symmetric theory. The operator in brackets (in the final expression) is also called the total derivative operator (with respect to t).

For example, the total derivative of $f(x(t), y(t))$ is

$$\frac{\mathrm{d}f}{\mathrm{d}t}=\frac{\partial f}{\partial x}\frac{\mathrm{d}x}{\mathrm{d}t}+\frac{\partial f}{\partial y}\frac{\mathrm{d}y}{\mathrm{d}t}.$$

Here there is no $\partial f / \partial t$ term since f itself does not depend on the independent variable t directly.

The Total Derivative via Differentials

Differentials provide a simple way to understand the total derivative. For instance, suppose $M(t,p_1,\ldots,p_n)$ is a function of time t and n variables p_i as in the previous section. Then, the differential of M is

$$\mathrm{d}M=\frac{\partial M}{\partial t}\mathrm{d}t+\sum_{i=1}^{n}\frac{\partial M}{\partial p_i}\mathrm{d}p_i.$$

This expression is often interpreted *heuristically* as a relation between infinitesimals. However, if the variables t and p_j are interpreted as functions $M(t,p_1,\ldots,p_n)$, and is interpreted to mean the composite of M with these functions, then the above expression makes perfect sense as an equality of differential 1-forms, and is immediate from the chain rule for the exterior derivative. The advantage of this point of view is that it takes into account arbitrary dependencies between the variables. For example, if $p_1^2=p_2p_3$ then $2p_1\,\mathrm{d}p_1=p_3\,\mathrm{d}p_2+p_2\,\mathrm{d}p_3$. In particular, if the variables p_j are all functions of t, as in the previous section, then

$$\mathrm{d}M=\frac{\partial M}{\partial t}\mathrm{d}t+\sum_{i=1}^{n}\frac{\partial M}{\partial p_i}\frac{\partial p_i}{\partial t}\,\mathrm{d}t.$$

The Total Derivative as a Linear Map

Let $U \subseteq \mathbb{R}^n$ be an open subset. Then a function $f : U \to \mathbb{R}^m$ is said to be (totally) differentiable at a point $p \in U$, if there exists a linear map $\mathrm{d}f_p : \mathbb{R}^n \to \mathbb{R}^m$ (also denoted $D_p f$ or $Df(p)$) such that

$$\lim_{x \to p} \frac{\| f(x) - f(p) - \mathrm{d}f_p(x-p) \|}{\| x - p \|} = 0.$$

The linear map $\mathrm{d}f_p$ is called the (total) derivative or (total) differential of f at p. A function is (totally) differentiable if its total derivative exists at every point in its domain.

Note that f is differentiable if and only if each of its components $f_i : U \to \mathbb{R}$ is differentiable. For this it is necessary, but not sufficient, that the partial derivatives of each function f_j exist. However, if these partial derivatives exist and are continuous, then f is differentiable and its differential at any point is the linear map determined by the Jacobian matrix of partial derivatives at that point.

Total Differential Equation

A *total differential equation* is a differential equation expressed in terms of total derivatives. Since the exterior derivative is a natural operator, in a sense that can be given a technical meaning, such equations are intrinsic and *geometric.*

Application of the Total Differential to Error Estimation

In measurement, the total differential is used in estimating the error Δf of a function f based on the errors Δx, Δy, ... of the parametres x, y, ... Assuming that the interval is short enough for the change to be approximately linear:

$$\Delta f(x) = f'(x) \times \Delta x$$

and that all variables are independent, then for all variables,

$$\Delta f = f_x \Delta x + f_y \Delta y + \cdots$$

This is because the derivative f_x with respect to the particular parameter x gives the sensitivity of the function f to a change in x, in particular the error Δx. As they are assumed to be independent, the analysis describes the worst-case scenario. The absolute values of the component errors are used, because after simple computation, the derivative may have a negative sign. From this principle the error rules of summation, multiplication etc. are derived, e.g.:

Let $f(a, b) = a \times b$;

$\Delta f = f_a \Delta a + f_b \Delta b$; evaluating the derivatives

$\Delta f = b\Delta a + a\Delta b$; dividing by f, which is $a \times b$

$\Delta f/f = \Delta a/a + \Delta b/b$

That is to say, in multiplication, the total relative error is the sum of the relative errors of the parameters.

Partial Derivative

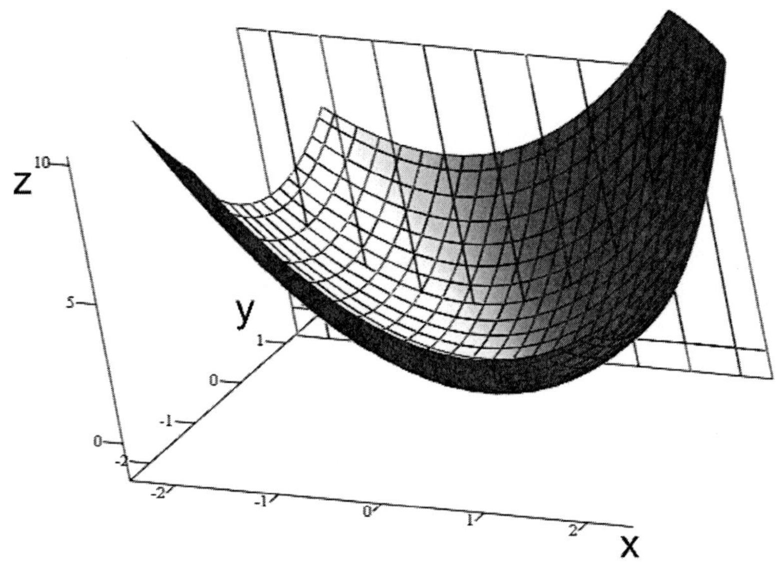

Figure: *A graph of $z = x^2 + xy + y^2$. For the partial derivative at (1, 1, 3) that leaves* y *constant, the corresponding tangent line is parallel to the* xz*-plane.*

In mathematics, a partial derivative of a function of several variables is its derivative with respect to one of those variables, with the others held constant (as opposed to the total derivative, in which all variables are allowed to vary). Partial derivatives are used in vector calculus and differential geometry.

The partial derivative of a function f with respect to the variable x is variously denoted by

$$f_x', f_{,x}, \partial_x f, \text{ or } \frac{\partial f}{\partial x}$$

The partial-derivative symbol is ∂. One of the first known uses of the symbol in mathematics is by Marquis de Condorcet from 1770, who used it for partial differences. The modern partial derivative notation is by Adrien-Marie Legendre (1786), though he later abandoned it; Carl Gustav Jacob Jacobi re-introduced the symbol in 1841.

Suppose that f is a function of more than one variable. For instance, $z = f(x,y) = x^2 + xy + y^2$.

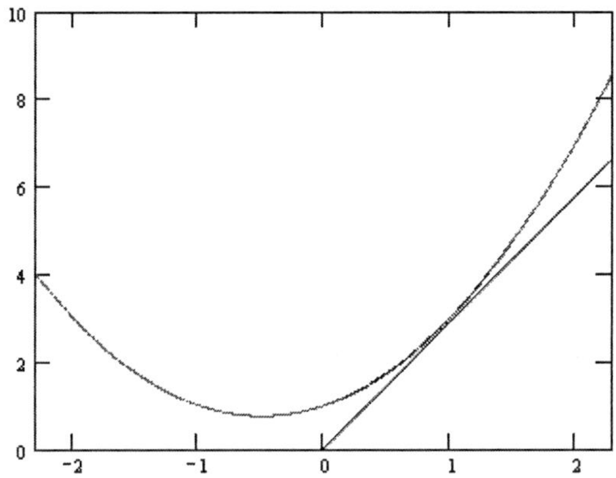

Figure: *A slice of the graph above at* y= *1*

The graph of this function defines a surface in Euclidean space. To every point on this surface, there is an infinite number of tangent lines. Partial differentiation is the act of choosing one of these lines and finding its slope.

Usually, the lines of most interest are those that are parallel to the xz-plane, and those that are parallel to the yz-plane (which result from holding either y or x constant, respectively.) To find the slope of the line tangent to the function at P(1, 1, 3) that is parallel to the xz-plane, the y variable is treated as constant. The graph and this plane are shown on the right. On the graph below it, we see the way the function looks on the plane $y = 1$. By finding the derivative of the equation while assuming that y is a constant, the slope of f at the point (x, y, z) is found to be:

$$\frac{\partial z}{\partial x} = 2x + y$$

So at (1, 1, 3), by substitution, the slope is 3. Therefore

$$\frac{\partial z}{\partial x} = 3$$

at the point. (1, 1, 3). That is, the partial derivative of z with respect to x at (1, 1, 3) is 3.

Definition

Basic Definition: The function f can be reinterpreted as a family of functions of one variable indexed by the other variables:

$$f(x,y) = f_x(y) = x^2 + xy + y^2.$$

In other words, every value of x defines a function, denoted f_x, which is a function of one variable. That is,

$$f_x(y) = x^2 + xy + y^2.$$

Once a value of x is chosen, say a, then $f(x,y)$ determines a function f_a which sends y to $a^2 + ay + y^2$:

$$f_a(y) = a^2 + ay + y^2.$$

In this expression, a is a *constant*, not a *variable*, so f_a is a function of only one real variable, that being y. Consequently, the definition of the derivative for a function of one variable applies:

$$f_{a'}(y) = a + 2y.$$

The above procedure can be performed for any choice of a. Assembling the derivatives together into a function gives a function which describes the variation of f in the y direction:

$$\frac{\partial f}{\partial y}(x,y) = x + 2y.$$

This is the partial derivative of f with respect to y. Here ∂ is a rounded d called the partial derivative symbol. To distinguish it from the letter d, ∂ is sometimes pronounced "del" or "partial" instead of "dee".

In general, the partial derivative of a function $f(x_1,...,x_n)$ in the direction x_i at the point $(a_1,...,a_n)$ is defined to be:

$$\frac{\partial f}{\partial x_i}(a_1,\ldots,a_n) = \lim_{h\to 0}\frac{f(a_1,\ldots,a_i+h,\ldots,a_n) - f(a_1,\ldots,a_i,\ldots,a_n)}{h}.$$

In the above difference quotient, all the variables except x_i are held fixed. That choice of fixed values determines a function of one variable $f_{a_1,\ldots,a_{i-1},a_{i+1},\ldots,a_n}(x_i) = f(a_1,\ldots,a_{i-1},x_i,a_{i+1},\ldots,a_n)$, and by definition,

$$\frac{df_{a_1,\ldots,a_{i-1},a_{i+1},\ldots,a_n}}{dx_i}(x_i) = \frac{\partial f}{\partial x_i}(a_1,\ldots,a_n).$$

In other words, the different choices of a index a family of one-variable functions just as in the example above. This expression also shows that the computation of partial derivatives reduces to the computation of one-variable derivatives.

An important example of a function of several variables is the case of a scalar-valued function $f(x_1,...x_n)$ on a domain in Euclidean space R^n (e.g., on R^2 or R^3). In this case f has a partial derivative $\partial f/\partial x_j$ with respect to each variable x_j. At the point a, these partial derivatives define the vector

$$\nabla f(a) = \left(\frac{\partial f}{\partial x_1}(a), \ldots, \frac{\partial f}{\partial x_n}(a)\right).$$

This vector is called the gradient of f at a. If f is differentiable at every point in some domain, then the gradient is a vector-valued function ∇f which takes the point a to the vector $\nabla f(a)$. Consequently, the gradient produces a vector field.

A common abuse of notation is to define the del operator (∇) as follows in three-dimensional Euclidean space R^3 with unit vectors $\hat{i}, \hat{j}, \hat{k}$:

$$\nabla = [\frac{\partial}{\partial x}]\hat{i} + [\frac{\partial}{\partial y}]\hat{j} + [\frac{\partial}{\partial z}]\hat{k}$$

Or, more generally, for n-dimensional Euclidean space R^n with coordinates $(x_1, x_2, x_3, \ldots, x_n)$ and unit vectors $(\hat{e}_1, \hat{e}_2, \hat{e}_3, \ldots, \hat{e}_n)$:

$$\nabla = \sum_{j=1}^{n}[\frac{\partial}{\partial x_j}]\hat{e}_j = [\frac{\partial}{\partial x_1}]\hat{e}_1 + [\frac{\partial}{\partial x_2}]\hat{e}_2 + [\frac{\partial}{\partial x_3}]\hat{e}_3 + \ldots + [\frac{\partial}{\partial x_n}]\hat{e}_n$$

Formal Definition

Like ordinary derivatives, the partial derivative is defined as a limit. Let U be an open subset of R^n and $f: U \to R$ a function. The partial derivative of f at the point $a = (a_1, \ldots, a_n) \in U$ with respect to the i-th variable a_i is defined as

$$\frac{\partial}{\partial a_i} f(a) = \lim_{h\to 0} \frac{f(a_1,\ldots,a_{i-1},a_i+h,a_{i+1},\ldots,a_n) - f(a_1,\ldots,a_i,\ldots,a_n)}{h}$$

Even if all partial derivatives $\partial f/\partial a_i(a)$ exist at a given point a, the function need not be continuous there. However, if all partial derivatives exist in a neighbourhood of a and are continuous there, then f is totally differentiable in that neighbourhood and the total derivative is continuous. In this case, it is said that f is a C^1 function. This can be used to generalize for vector valued functions ($f : U \to R^m$) by carefully using a componentwise argument.

The partial derivative $\frac{\partial f}{\partial x}$ can be seen as another function defined on U and can again be partially differentiated. If all mixed second order partial derivatives are continuous at a point (or on a set), f is termed a C^2 function at that point (or on that set); in this case, the partial derivatives can be exchanged by Clairaut's theorem:

$$\frac{\partial^2 f}{\partial x_i \partial x_j} = \frac{\partial^2 f}{\partial x_j \partial x_i}.$$

Examples

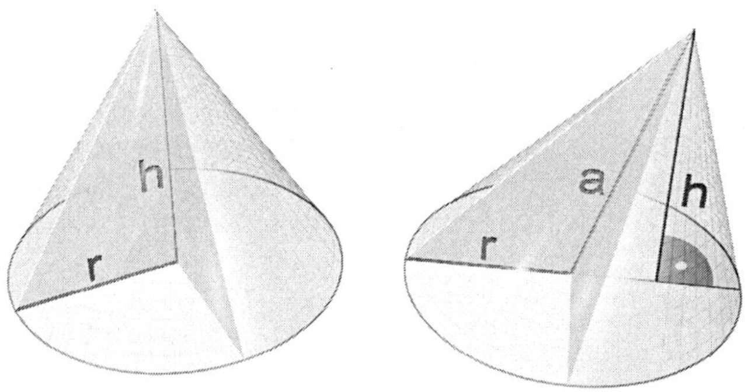

Figure: *The volume of a cone depends on height and radius*

The volume V of a cone depends on the cone's height h and its radius r according to the formula

$$V(r,h) = \frac{\pi r^2 h}{3}.$$

The partial derivative of V with respect to r is

$$\frac{\partial V}{\partial r} = \frac{2\pi r h}{3},$$

which represents the rate with which a cone's volume changes if its radius is varied and its height is kept constant. The partial derivative with respect to h is

$$\frac{\partial V}{\partial h} = \frac{\pi r^2}{3},$$

which represents the rate with which the volume changes if its height is varied and its radius is kept constant.

By contrast, the *total* derivative of V with respect to r and h are respectively

$$\frac{\mathrm{d}V}{\mathrm{d}r} = \overbrace{\frac{2\pi rh}{3}}^{\frac{\partial V}{\partial r}} + \overbrace{\frac{\pi r^2}{3}}^{\frac{\partial V}{\partial h}} \frac{\mathrm{d}h}{\mathrm{d}r}$$

and

$$\frac{\mathrm{d}V}{\mathrm{d}h} = \overbrace{\frac{\pi r^2}{3}}^{\frac{\partial V}{\partial h}} + \overbrace{\frac{2\pi rh}{3}}^{\frac{\partial V}{\partial r}} \frac{\mathrm{d}r}{\mathrm{d}h}$$

The difference between the total and partial derivative is the elimination of indirect dependencies between variables in partial derivatives.

If (for some arbitrary reason) the cone's proportions have to stay the same, and the height and radius are in a fixed ratio k,

$$k = \frac{h}{r} = \frac{\mathrm{d}h}{\mathrm{d}r}.$$

This gives the total derivative with respect to r:

$$\frac{\mathrm{d}V}{\mathrm{d}r} = \frac{2\pi rh}{3} + \frac{\pi r^2}{3}k$$

Which simplifies to:

$$\frac{\mathrm{d}V}{\mathrm{d}r} = k\pi r^2$$

Similarly, the total derivative with respect to h is:

$$\frac{\mathrm{d}V}{\mathrm{d}h} = \pi r^2$$

Equations involving an unknown function's partial derivatives are called partial differential equations and are common in physics, engineering, and other sciences and applied disciplines.

Notation

For the following examples, let f be a function in x, y and z.

First-order partial derivatives:

$$\frac{\partial f}{\partial x} = f_x = \partial_x f.$$

Second-order partial derivatives:

$$\frac{\partial^2 f}{\partial x^2} = f_{xx} = \partial_{xx} f.$$

Second-order mixed derivatives:

$$\frac{\partial^2 f}{\partial y \partial x} = \frac{\partial}{\partial y}\left(\frac{\partial f}{\partial x}\right) = (f_x)_y = f_{xy} = \partial_{yx} f.$$

Higher-order partial and mixed derivatives:

$$\frac{\partial^{i+j+k} f}{\partial x^i \partial y^j \partial z^k} = f^{(i,j,k)}.$$

When dealing with functions of multiple variables, some of these variables may be related to each other, and it may be necessary to specify explicitly which variables are being held constant. In fields such as statistical mechanics, the partial derivative of f with respect to x, holding y and z constant, is often expressed as

$$\left(\frac{\partial f}{\partial x}\right)_{y,z}.$$

Antiderivative Analogue

There is a concept for partial derivatives that is analogous to antiderivatives for regular derivatives. Given a partial derivative, it allows for the partial recovery of the original function.

Consider the example of $\frac{\partial z}{\partial x} = 2x + y$. The "partial" integral can be taken with respect to x (treating y as constant, in a similar manner to partial derivation):

$$z = \int \frac{\partial z}{\partial x} dx = x^2 + xy + g(y)$$

Here, the "constant" of integration is no longer a constant, but instead a function of all the variables of the original function except x. The reason for this is that all the other variables are treated as constant when taking the partial derivative, so any function which does not involve x will disappear when taking the partial derivative, and we have to account for this when we take the antiderivative.

The most general way to represent this is to have the "constant" represent an unknown function of all the other variables.

Thus the set of functions $x^2 + xy + g(y)$, where g is any one-argument function, represents the entire set of functions in variables x,y that could have produced the x-partial derivative $2x+y$.

If all the partial derivatives of a function are known (for example, with the gradient), then the antiderivatives can be matched via the above process to reconstruct the original function up to a constant.

Symmetry of Second Derivatives

In mathematics, the symmetry of second derivatives (also called the equality of mixed partials) refers to the possibility of interchanging the order of taking partial derivatives of a function

$$f(x_1, x_2, \ldots, x_n)$$

of n variables. If the partial derivative with respect to x_i is denoted with a subscript i, then the symmetry is the assertion that the second-order partial derivatives f_{ij} satisfy the identity

$$f_{ij} = f_{ji}$$

so that they form an $n \times n$ symmetric matrix. This is sometimes known as Young's theorem.

In the context of partial differential equations it is called the Schwarz integrability condition.

Hessian Matrix

This matrix of second-order partial derivatives of f is called the Hessian matrix of f. The entries in it off the main diagonal are the mixed derivatives; that is, successive partial derivatives with respect to different variables.

In most "real-life" circumstances the Hessian matrix is symmetric, although there are- mathematically- a far greater number of functions that do not have this property. Mathematical analysis reveals that symmetry requires a hypothesis on f that goes further than simply stating the existence of the second derivatives at a particular point. Clairaut's theorem gives a sufficient condition on f for this to occur.

Formal Expressions of Symmetry

In symbols, the symmetry says that, for example,

$$\frac{\partial}{\partial x}\left(\frac{\partial f}{\partial y}\right) = \frac{\partial}{\partial y}\left(\frac{\partial f}{\partial x}\right).$$

This equality can also be written as

$$\partial_{xy} f = \partial_{yx} f.$$

Alternatively, the symmetry can be written as an algebraic statement involving the differential operator D_i which takes the partial derivative with respect to x_i:

$$D_i \cdot D_j = D_j \cdot D_i.$$

From this relation it follows that the ring of differential operators with constant coefficients, generated by the D_i, is commutative. But one should naturally specify some domain for these operators. It is easy to check the symmetry as applied to monomials, so that one can take polynomials in the x_i as a domain. In fact smooth functions are possible.

Clairaut's Theorem

In mathematical analysis, *Clairaut's theorem* (or *Schwarz's theorem*) named after Alexis Clairaut and Hermann Schwarz, states that if

$$f : \mathbb{R}^n \to \mathbb{R}$$

has continuous second partial derivatives at any given point in $\mathbb{R}^n$, say, $(a_1,\ldots,a_n)$, then $\forall i,j \in \mathbb{N} \setminus \{0\} : i,j \le n$,

$$\frac{\partial^2 f}{\partial x_i \partial x_j}(a_1,\ldots,a_n) = \frac{\partial^2 f}{\partial x_j \partial x_i}(a_1,\ldots,a_n).$$

The partial derivations of this function are commutative at that point. One easy way to establish this theorem (in the case where $n = 2$, $i = 1$, and $j = 2$, which readily entails the result in general) is by applying Green's theorem to the gradient of f.

Distribution Theory Formulation

The theory of distributions eliminates analytic problems with the symmetry. The derivative of any integrable function can be defined as a distribution. The use of integration by parts puts the symmetry question back onto the test functions, which are smooth and certainly satisfy the symmetry. In the sense of distributions, symmetry always holds.

Another approach, which defines the Fourier transform of a function, is to note that on such transforms partial derivatives become multiplication operators that commute much more obviously.

Requirement of Continuity

The symmetry may be broken if the function fails to satisfy the premises of Clairaut's theorem, such as if the derivatives are not continuous.

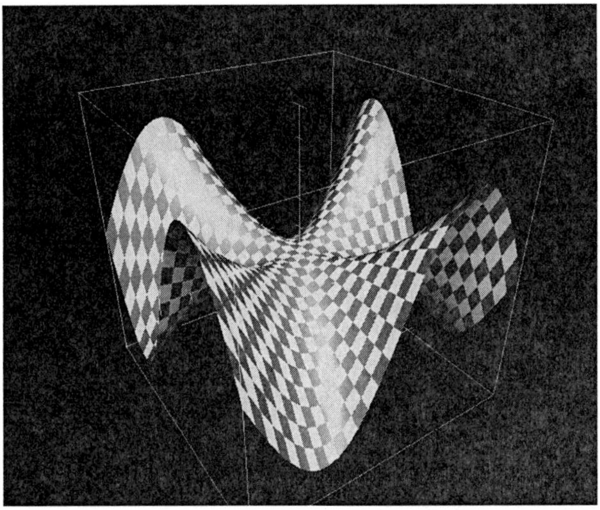

Figure: *This function,* f(x,y), *does not have symmetric second derivatives at its origin.*

An example of non-symmetry is the function:

$$f(x,y) = \begin{cases} \dfrac{xy(x^2 - y^2)}{x^2 + y^2} & \text{for } (x,y) \neq (0,0) \\ 0 & \text{for } (x,y) = (0,0). \end{cases}$$

Although this function is everywhere continuous, its algebraic derivatives are undefined at the origin point. Elsewhere along the x-axis the y-derivative $\partial_y f\,|_{(x,0)} = x$, and so:

$$\partial_x \partial_y f\,|_{(0,0)} = \lim_{\epsilon \to 0} \frac{\partial_y f\,|_{(\epsilon,0)} - \partial_y f\,|_{(0,0)}}{\epsilon} = 1.$$

Vice versa, along the y-axis the x-derivative $\partial_x f\,|_{(0,y)} = -y$, and so $\partial_y \partial_x f\,|_{(0,0)} = -1$. That is, $\partial_{xy} f \neq \partial_{yx} f$ at (0, 0), although the mixed partial derivates of this function do exist and are symmetric at every other point.

In general, the interchange of limiting operations need not commute. Given two variables near (0, 0) and two limiting processes on

$$f(h,k) - f(h,0) - f(0,k) + f(0,0)$$

corresponding to making $h \to 0$ first, and to making $k \to 0$ first. It can matter, looking at the first-order terms, which is applied first. This leads to the construction of pathological examples in which second derivatives are non-symmetric.

This kind of example belongs to the theory of real analysis where the pointwise value of functions matters. When viewed as a distribution the second partial derivative's values can be changed at an arbitrary set of points as long as this has Lebesgue measure 0. Since in the example the Hessian is symmetric everywhere except (0,0), there is no contradiction with the fact that the Hessian, viewed as a Schwartz distribution, is symmetric.

In Lie Theory

Consider the first-order differential operators D_i to be infinitesimal operators on Euclidean space. That is, D_i in a sense generates the one-parameter group of translations parallel to the x_i-axis. These groups commute with each other, and therefore the infinitesimal generators do also; the Lie bracket

$$[D_i, D_j] = 0$$

is this property's reflection. In other words, the Lie derivative of one coordinate with respect to another is zero.

Homogeneous Function

In mathematics, a homogeneous function is a function with multiplicative scaling behaviour: if the argument is multiplied by a factor, then the result is multiplied by some power of this factor. More precisely, if $f : V \to W$ is a function between two vector spaces over a field F, and k is an integer, then f is said to be homogeneous of degree k if

$$f(\alpha \mathrm{v}) = \alpha^k f(\mathrm{v}) \qquad (1)$$

for all nonzero $\alpha \in F$ and $\mathrm{v} \in V$. This implies it has scale invariance. When the vector spaces involved are over the real numbers, a slightly more general form of homogeneity is often used, requiring only that (1) hold for all $\alpha > 0$.

Homogeneous functions can also be defined for vector spaces with the origin deleted, a fact that is used in the definition of sheaves on projective space in algebraic geometry. More generally, if $S \subset V$ is any subset that is invariant under scalar multiplication by elements of the field (a "cone"), then an homogeneous function from S to W can still be defined by (1).

Examples

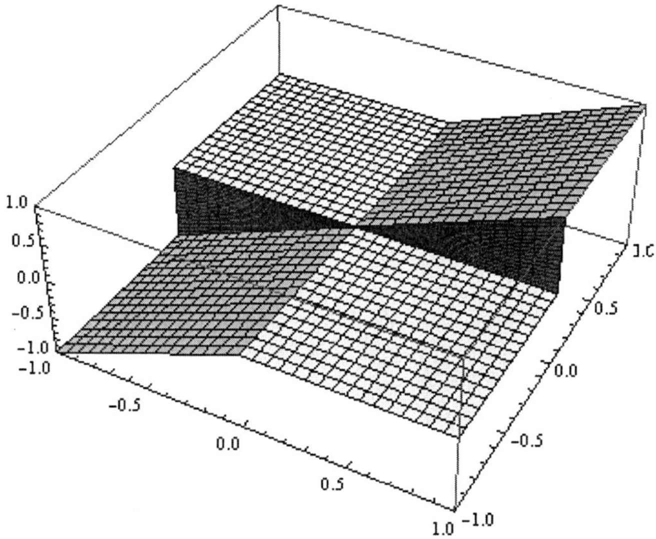

Figure: *A homogeneous function is not necessarily continuous, as shown by this example. This is the function* f *defined by* $f(x,y) = x$ *if* $xy > 0$ *or* $f(x,y) = 0$ *if* $xy \leq 0$. *This function is homogeneous of order 1, i.e.* $f(\alpha x, \alpha y) = \alpha f(x,y)$ *for any real numbers* α, x, y. *It is discontinuous at* $y = 0$.

Linear Functions

Any Linear Function f : $V \to W$ is homogeneous of degree 1, since by the definition of linearity $f(\alpha \mathrm{v}) = \alpha f(\mathrm{v})$ for all $\alpha \in F$ and v $\in V$. Similarly, any multilinear function $f : V_1 \times V_2 \times \ldots V_n \to W$ is homogeneous of degree n, since by the definition of multilinearity

$$f(\alpha \mathrm{v}_1, \ldots, \alpha \mathrm{v}_n) = \alpha^n f(\mathrm{v}_1, \ldots, \mathrm{v}_n)$$

for all $\alpha \in F$ and $\mathrm{v}_1 \in V_1$, $\mathrm{v}_2 \in V_2$, ..., $\mathrm{v}_n \in V_n$. It follows that the n-th differential of a function $f : X \to Y$ between two Banach spaces X and Y is homogeneous of degree n.

Homogeneous Polynomials

Monomials in n variables define homogeneous functions $f : F^n \to F$. For example,

$$f(x,y,z) = x^5 y^2 z^3$$

is homogeneous of degree 10 since

$$f(\alpha x, \alpha y, \alpha z) = (\alpha x)^5 (\alpha y)^2 (\alpha z)^3 = \alpha^{10} x^5 y^2 z^3 = \alpha^{10} f(x,y,z).$$

The degree is the sum of the exponents on the variables; in this example, 10=5+2+3.

A homogeneous polynomial is a polynomial made up of a sum of monomials of the same degree. For example,

$$x^5 + 2x^3y^2 + 9xy^4$$

is a homogeneous polynomial of degree 5. Homogeneous polynomials also define homogeneous functions.

Polarization

A multilinear function $g: V \times V \times \ldots V \to F$ from the n-th Cartesian product of V with itself to the underlying field F gives rise to an homogeneous function $f : V \to F$ by evaluating on the diagonal:

$$f(v) = g(v,v,\ldots,v).$$

The resulting function f is a polynomial on the vector space V.

Conversely, if F has characteristic zero, then given an homogeneous polynomial f of degree n on V, the polarization of f is a multilinear function $g : V\times V\times \ldots V\to F$ on the n-th Cartesian product of V. The polarization is defined by

$$g(v_1,v_2,\ldots,v_n) = \frac{1}{n!}\frac{\partial}{\partial t_1}\frac{\partial}{\partial t_2}\cdots\frac{\partial}{\partial t_n} f(t_1v_1 + \cdots + t_nv_n).$$

These two constructions, one of an homogeneous polynomial from a multilinear form and the other of a multilinear form from an homogeneous polynomial, are mutually inverse to one another. In finite dimensions, they establish an isomorphism of graded vector spaces from the symmetric algebra of V' to the algebra of homogeneous polynomials on V.

Rational Functions

Rational functions formed as the ratio of two *homogeneous* polynomials are homogeneous functions off of the affine cone cut out by the zero locus of the denominator. Thus, if f is homogeneous of degree m and g is homogeneous of degree n, then f/g is homogeneous of degree $m - n$ away from the zeros of g.

Non-examples

Logarithms: The natural logarithm $f(x) = \ln x$ scales additively and so is not homogeneous. This can be proved by noting that $f(5x) = \ln 5x = \ln 5 + f(x)$, $f(10x) = \ln 10 + f(x)$, and $f(15x) = \ln 15 + f(x)$. Therefore $\nexists\, k$ such that $f(\alpha\cdot x) = \alpha^k\cdot f(x)\cdot$

Affine Functions

Affine functions (the function $f(x) = x + 5$ is an example) do not scale multiplicatively.

Positive Homogeneity

In the special case of vector spaces over the real numbers, the notation of positive homogeneity often plays a more important role than homogeneity in the above sense. A function $f : V \setminus \{0\} \rightarrow \mathrm{R}$ is positive homogeneous of degree k if

$$f(\alpha x) = \alpha^k f(x)$$

for all $\alpha > 0$. Here k can be any complex number. A (nonzero) continuous function homogeneous of degree k on $\mathrm{R}^n \setminus \{0\}$ extends continuously to R^n if and only if $\mathrm{Re}\{k\} > 0$.

Positive homogeneous functions are characterized by Euler's homogeneous function theorem. Suppose that the function $f : \mathrm{R}^n \setminus \{0\} \rightarrow \mathrm{R}$ is continuously differentiable. Then f is positive homogeneous of degree k if and only if

$$\mathrm{x}\cdot\nabla f(\mathrm{x}) = kf(\mathrm{x}).$$

This result follows at once by differentiating both sides of the equation $f(\alpha \mathrm{y}) = \alpha^k f(\mathrm{y})$ with respect to α, applying the chain rule, and choosing α to be 1. The converse holds by integrating. Specifically, let $g(\alpha) = f(\alpha \mathrm{x})$. Since $\alpha \mathrm{x}\cdot\nabla f(\alpha \mathrm{x}) = kf(\alpha \mathrm{x})$,

$$g'(\alpha) = \mathrm{x}\cdot\nabla f(\alpha \mathrm{x}) = \frac{k}{\alpha} f(\alpha \mathrm{x}) = \frac{k}{\alpha} g(\alpha).$$

Thus, $g'(\alpha) - \frac{k}{\alpha} g(\alpha) = 0$. This implies $g(\alpha) = g(1)\alpha^k$. Therefore, $f(\alpha \mathrm{x}) = g(\alpha) = \alpha^k g(1) = \alpha^k f(\mathrm{x})$: f is positive homogeneous of degree k.

As a consequence, suppose that $f : \mathrm{R}^n \rightarrow \mathrm{R}$ is differentiable and homogeneous of degree k. Then its first-order partial derivatives $\partial f / \partial x_i$ are homogeneous of degree $k - 1$. The result follows from Euler's theorem by commuting the operator $\mathrm{x}\cdot\nabla$ with the partial derivative.

Homogeneous Distributions

A compactly supported continuous function f on R^n is homogeneous of degree k if and only if

$$\int_{\mathbb{R}^n} f(tx)\varphi(x)dx = t^k \int_{\mathbb{R}^n} f(x)\varphi(x)dx$$

for all compactly supported test functions φ; and nonzero real t. Equivalently, making a change of variable $y = tx$, f is homogeneous of degree k if and only if

$$t^{-n}\int_{\mathbb{R}^n} f(y)\varphi(y/t)dy = t^k \int_{\mathbb{R}^n} f(y)\varphi(y)dy$$

for all t and all test functions φ;. The last display makes it possible to define homogeneity of distributions. A distribution S is homogeneous of degree k if

$$t^{-n}\langle S, \varphi^\circ\mu_t\rangle = t^k\langle S, \varphi\rangle$$

for all nonzero real t and all test functions φ;. Here the angle brackets denote the pairing between distributions and test functions, and $\mu_t : \mathrm{R}^n \to \mathrm{R}^n$ is the mapping of scalar multiplication by the real number t.

Application to Differential Equations

The substitution $v = y/x$ converts the ordinary differential equation

$$I(x,y)\frac{\mathrm{d}y}{\mathrm{d}x} + J(x,y) = 0,$$

where I and J are homogeneous functions of the same degree, into the separable differential equation

$$x\frac{\mathrm{d}v}{\mathrm{d}x} = -\frac{J(1,v)}{I(1,v)} - v.$$

Maxima and Minima

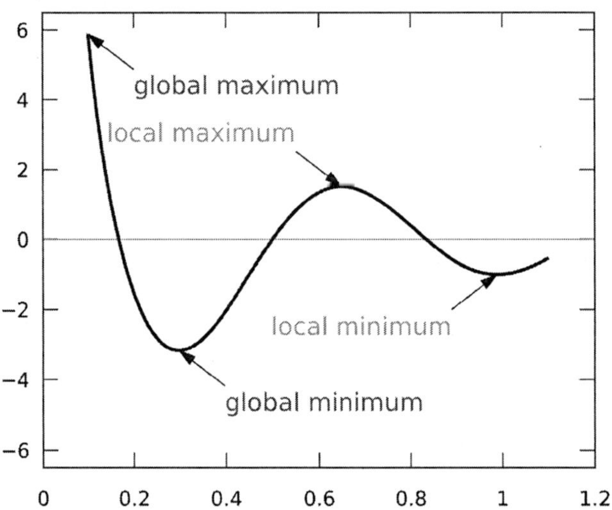

Figure: *Local and global maxima and minima for* $\cos(3\pi x)/x$, $0.1 \le x \le 1.1$

In mathematics, the maximum and minimum (plural: maxima and minima) of a function, known collectively as extrema (singular: extremum), are the largest and smallest value that the function takes at a point either within a given neighbourhood (*local* or *relative* extremum) or on the function domain in its entirety (*global* or *absolute* extremum). Pierre de Fermat was one of the first mathematicians to propose a general technique (called adequality) for finding maxima and minima.

More generally, the maximum and minimum of a set (as defined in set theory) are the greatest and least element in the set. Unbounded infinite sets such as the set of real numbers have no minimum and maximum.

To locate extreme values is the basic objective of optimization.

Analytical Definition

A real-valued function f defined on a real line is said to have a local (or relative) maximum point at the point x^*, if there exists some $\varepsilon > 0$ such that $f(x^*) \geq f(x)$ when $|x - x^*| < \varepsilon$. The value of the function at this point is called maximum of the function. Similarly, a function has a local minimum point at x^*, if $f(x^*) \leq f(x)$ when $|x - x^*| < \varepsilon$. The value of the function at this point is called minimum of the function.

A function has a global (or absolute) maximum point at x^* if $f(x^*) \leq f(x)$ for all x. Similarly, a function has a global (or absolute) minimum point at x^* if $f(x^*) \leq f(x)$ for all x. The global maximum and global minimum points are also known as the arg max and arg min: the argument (input) at which the maximum (respectively, minimum) occurs.

Restricted Domains: There may be maxima and minima for a function whose domain does not include all real numbers. A real-valued function, whose domain is any set, can have a global maximum and minimum. There may also be local maxima and local minima points, but only at points of the domain set where the concept of neighbourhood is defined. A neighbourhood plays the role of the set of x such that $|x - x^*| < \varepsilon$.

A continuous (real-valued) function on a compact set always takes maximum and minimum values on that set. An important example is a function whose domain is a closed (and bounded) interval of real numbers. The neighbourhood requirement precludes a *local* maximum or minimum at an endpoint of an interval. However, an endpoint may still be a *global* maximum or minimum. Thus it is *not always true,* for finite domains, that a global maximum (minimum) must also be a local maximum (minimum).

Finding Functional Maxima and Minima

Finding global maxima and minima is the goal of mathematical optimization. If a function is continuous on a closed interval, then by the extreme value theorem global maxima and minima exist. Furthermore, a global maximum (or minimum) either must be a local maximum (or minimum) in the interior of the domain, or must lie on the boundary of the domain. So a method of finding a global maximum (or minimum) is to look at all the local maxima (or minima) in the interior, and also look at the maxima (or minima) of the points on the boundary; and take the biggest (or smallest) one.

Local extrema can be found by Fermat's theorem, which states that they must occur at critical points. One can distinguish whether a critical point is a local maximum or local minimum by using the first derivative test or second derivative test.

For any function that is defined piecewise, one finds a maximum (or minimum) by finding the maximum (or minimum) of each piece separately; and then seeing which one is biggest (or smallest).

Examples

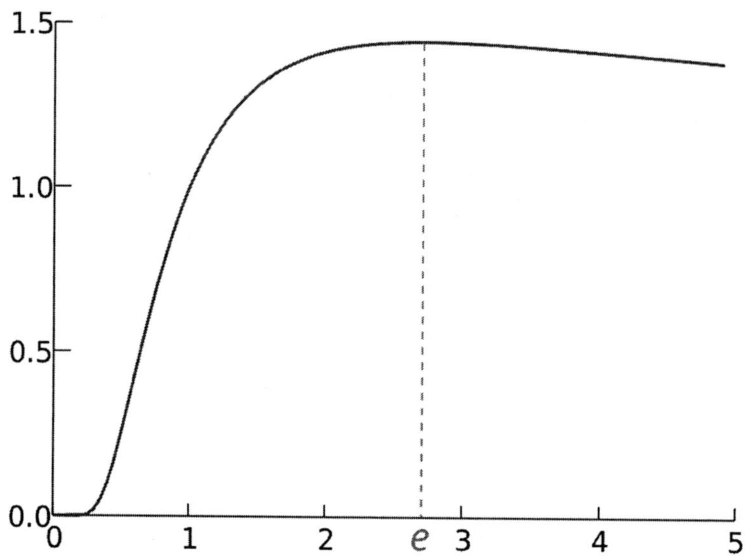

Figure: *The global maximum of* $\sqrt[x]{x}$ *occurs at* x = e.

- The function x^2 has a unique global minimum at $x = 0$.
- The function x^3 has no global minima or maxima. Although the first derivative ($3x^2$) is 0 at $x = 0$, this is an inflection point.
- The function $\sqrt[x]{x}$ has a unique global maximum at $x = e$.
- The function x^{-x} has a unique global maximum over the positive real numbers at $x = 1/e$.
- The function $x^3/3 - x$ has first derivative $x^2 - 1$ and second derivative $2x$. Setting the first derivative to 0 and solving for x gives stationary points at –1 and +1. From the sign of the second derivative we can see that –1 is a local maximum and +1 is a local minimum. Note that this function has no global maximum or minimum.
- The function $|x|$ has a global minimum at $x = 0$ that cannot be found by taking derivatives, because the derivative does not exist at $x = 0$.
- The function $\cos(x)$ has infinitely many global maxima at 0, $\pm 2\pi$, $\pm 4\pi$, ..., and infinitely many global minima at $\pm\pi$, $\pm 3\pi$, ...
- The function $2\cos(x) - x$ has infinitely many local maxima and minima, but no global maximum or minimum.
- The function $\cos(3\pi x)/x$ with $0.1 \le x \le 1.1$ has a global maximum at $x = 0.1$ (a boundary), a global minimum near $x = 0.3$, a local maximum near $x = 0.6$, and a local minimum near $x = 1.0$.
- The function $x^3 + 3x^2 - 2x + 1$ defined over the closed interval (segment) [–4,2] has two extrema: one local maximum at $x = -1-{}^{\sqrt{15}}\text{D}_{3}$, one local minimum at $x = -1+{}^{\sqrt{15}}\text{D}_{3}$, a global maximum at $x = 2$ and a global minimum at $x = -4$.

Functions of More than one Variable

For functions of more than one variable, similar conditions apply. For example, in the (enlargeable) figure at the right, the necessary conditions for a *local* maximum are similar to those of a function with only one variable. The first partial derivatives as to z (the variable to be maximized) are zero at the maximum (the glowing dot on top in the figure). The second partial derivatives are negative. These are only necessary, not sufficient, conditions for a local maximum because of the possibility of a saddle point. For use of these conditions to solve for a maximum, the function z must also be differentiable throughout. The second partial derivative test can help classify the point as a relative maximum or relative minimum.

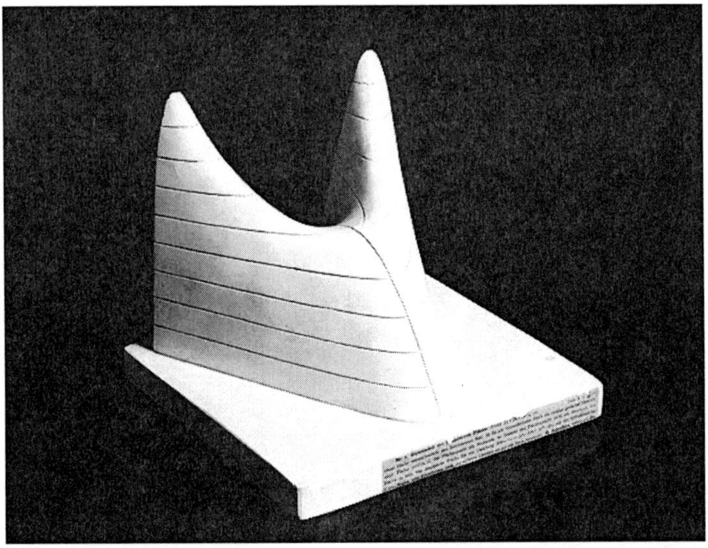

Figure: *Peano surface, a counterexample to some criteria of local maxima of the 19th century*

In contrast, there are substantial differences between functions of one variable and functions of more than one variable in the identification of global extrema. For example, if a bounded differentiable function f defined on a closed interval in the real line has a single critical point, which is a local minimum, then it is also a global minimum (use the intermediate value theorem and Rolle's theorem to prove this by reductio ad absurdum). In two and more dimensions, this argument fails, as the function

$$f(x,y) = x^2 + y^2(1-x)^3, \qquad x, y \in \mathbb{R},$$

shows. Its only critical point is at (0,0), which is a local minimum with $f(0,0) = 0$. However, it cannot be a global one, because $f(4,1) = -11$.

In Relation to Sets

Maxima and minima are more generally defined for sets. In general, if an ordered set S has a greatest element m, m is a maximal element. Furthermore, if S is a subset of an ordered set T and m is the greatest element of S with respect to order induced by T, m is a least upper bound of S in T. The similar result holds for least element, minimal element and greatest lower bound.

In the case of a general partial order, the least element (smaller than all other) should not be confused with a minimal element (nothing is smaller). Likewise, a greatest element of a partially ordered set (poset) is an upper bound of the set which is contained within the set, whereas a maximal element m of a poset A is an element of A such that if $m \leq b$ (for any b in A) then $m = b$. Any least element or greatest element of a poset is unique, but a poset can have several minimal or maximal elements. If a poset has more than one maximal element, then these elements will not be mutually comparable.

In a totally ordered set, or *chain*, all elements are mutually comparable, so such a set can have at most one minimal element and at most one maximal element. Then, due to mutual comparability, the minimal element will also be the least element and the maximal element will also be the greatest element. Thus in a totally ordered set we can simply use the terms *minimum* and *maximum*. If a chain is finite then it will always have a maximum and a minimum. If a chain is infinite then it need not have a maximum or a minimum. For example, the set of natural numbers has no maximum, though it has a minimum. If an infinite chain S is bounded, then the closure $Cl(S)$ of the set occasionally has a minimum and a maximum, in such case they are called the greatest lower bound and the least upper bound of the set S, respectively..

Integral

Integration is an important concept in mathematics and, together with its inverse, differentiation, is one of the two main operations in calculus. Given a function f of a real variable x and an interval $[a, b]$ of the real line, the definite integral

$$\int_a^b f(x)dx$$

is defined informally to be the signed area of the region in the xy-plane bounded by the graph of f, the x-axis, and the vertical lines $x = a$ and $x = b$, such that area above the x-axis adds to the total, and that below the x-axis subtracts from the total. The term *integral* may also refer to the notion of the antiderivative, a function F whose derivative is the given function f. In this case, it is called an *indefinite integral* and is written:

$$F = \int f(x)dx.$$

The integrals discussed in this article are termed *definite integrals*.

The principles of integration were formulated independently by Isaac Newton and Gottfried Leibniz in the late 17th century. Through the fundamental theorem of calculus, which they independently developed, integration is connected with differentiation: if f is a continuous real-valued function defined on a closed interval $[a, b]$, then, once an antiderivative F of f is known, the definite integral of f over that interval is given by

$$\int_a^b f(x)dx = F(b) - F(a)$$

Integrals and derivatives became the basic tools of calculus, with numerous applications in science and engineering. The founders of the calculus thought of the integral as an infinite sum of rectangles of infinitesimal width. A rigorous mathematical definition of the integral was given by Bernhard Riemann. It is based on a limiting procedure which approximates the area of a curvilinear region by breaking the region into thin vertical slabs. Beginning in the nineteenth century, more sophisticated notions of integrals began to appear, where the type of the function as well as the domain over which the integration is performed has been generalised. A line integral is defined for functions of two or three variables, and the interval of integration $[a, b]$ is replaced by a certain curve connecting two points on the plane or in the space. In a surface integral, the curve is replaced by a piece of a surface in the three-dimensional space. Integrals of differential forms play a fundamental role in modern differential geometry. These generalizations of integrals first arose from the needs of physics, and they play an important role in the formulation of many physical laws, notably those of electrodynamics. There are many modern concepts of integration, among these, the most common is based on the abstract mathematical theory known as Lebesgue integration, developed by Henri Lebesgue.

Pre-calculus Integration

The first documented systematic technique capable of determining integrals is the method of exhaustion of the ancient Greek astronomer Eudoxus (*ca.* 370 BC), which sought to find areas and volumes by breaking them up into an infinite number of shapes for which the area or volume was known. This method was further developed and employed by Archimedes in the 3rd century BC and used to calculate areas for parabolas and an approximation to the area of a circle. Similar methods were independently developed in China around the 3rd century AD by Liu Hui, who used it to find the area of the circle. This method was later used in the 5th century by Chinese father-and-son mathematicians Zu Chongzhi and Zu Geng to find the volume of a sphere (Shea 2007; Katz 2004, pp. 125–126).

The next significant advances in integral calculus did not begin to appear until the 16th century. At this time the work of Cavalieri with his *method of indivisibles*, and work by Fermat, began to lay the foundations of modern calculus, with Cavalieri computing the integrals of x^n up to degree $n = 9$ in Cavalieri's quadrature formula. Further steps were made in the early 17th century by Barrow and Torricelli, who provided the first hints of a connection between integration and differentiation. Barrow provided the first proof of the fundamental theorem of calculus. Wallis generalized Cavalieri's method, computing integrals of x to a general power, including negative powers and fractional powers.

Newton and Leibniz

The major advance in integration came in the 17th century with the independent discovery of the fundamental theorem of calculus by Newton and Leibniz. The theorem demonstrates a connection between integration and differentiation.

This connection, combined with the comparative ease of differentiation, can be exploited to calculate integrals. In particular, the fundamental theorem of calculus allows one to solve a much broader class of problems. Equal in importance is the comprehensive mathematical framework that both Newton and Leibniz developed. Given the name infinitesimal calculus, it allowed for precise analysis of functions within continuous domains. This framework eventually became modern calculus, whose notation for integrals is drawn directly from the work of Leibniz.

Formalizing Integrals

While Newton and Leibniz provided a systematic approach to integration, their work lacked a degree of rigour. Bishop Berkeley memorably attacked the vanishing increments used by Newton, calling them "ghosts of departed quantities". Calculus acquired a firmer footing with the development of limits. Integration was first rigorously formalized, using limits, by Riemann. Although all bounded piecewise continuous functions are Riemann integrable on a bounded interval, subsequently more general functions were considered – particularly in the context of Fourier analysis – to which Riemann's definition does not apply, and Lebesgue formulated a different definition of integral, founded in measure theory (a subfield of real analysis). Other definitions of integral, extending Riemann's and Lebesgue's approaches, were proposed. These approaches based on the real number system are the ones most common today, but alternative approaches exist, such as a definition of integral as the standard part of an infinite Riemann sum, based on the hyperreal number system.

Historical Notation

Isaac Newton used a small vertical bar above a variable to indicate integration, or placed the variable inside a box. The vertical bar was easily confused with $\dot{x}$ or x', which Newton used to indicate differentiation, and the box notation was difficult for printers to reproduce, so these notations were not widely adopted.

The modern notation for the indefinite integral was introduced by Gottfried Leibniz in 1675 (Burton 1988, p. 359; Leibniz 1899, p. 154). He adapted the integral symbol, ∫, from the letter ſ (long s), standing for *summa* (written as *numma*; Latin for "sum" or "total"). The modern notation for the definite integral, with limits above and below the integral sign, was first used by Joseph Fourier in *Mémoires* of the French Academy around 1819–20, reprinted in his book of 1822 (Cajori 1929, pp. 249–250; Fourier 1822, §231).

Terminology and Notation

The simplest case, the integral over x of a real-valued function $f(x)$, is written as

$$\int f(x)dx.$$

The integral sign ∫ represents integration. The dx indicates that we are integrating over x; x is called the variable of integration. In correct mathematical typography, the dx is separated from the integrand by a space (as shown). Some authors use an upright d (that is, dx instead of dx). Inside the $\int \ldots dx$ is the expression to be integrated, called the *integrand*. In this case the integrand is the function $f(x)$. Because there is no domain specified, the integral is called an *indefinite integral.*

When integrating over a specified domain, we speak of a *definite integral.* Integrating over a domain D is written as

$\int_D f(x)dx$, or $\int_a^b f(x)dxa$ if the domain is an interval $[a, b]$ of x;

The domain D or the interval $[a, b]$ is called the *domain of integration.* If a function has an integral, it is said to be *integrable.* In general, the integrand may be a function of more than one variable, and the domain of integration may be an area, volume, a higher dimensional region, or even an abstract space that does not have a geometric structure in any usual sense (such as a sample space in probability theory).

In the modern Arabic mathematical notation, which aims at pre-university levels of education in the Arab world and is written from right to left, a reflected integral symbol ⨚ is used (W3C 2006).

The variable of integration *dx* has different interpretations depending on the theory being used. It can be seen as strictly a notation indicating that *x* is a dummy variable of integration; if the integral is seen as a Riemann sum, *dx* is a reflection of the weights or widths *d* of the intervals of *x*; in Lebesgue integration and its extensions, *dx* is a measure; in non-standard analysis, it is an infinitesimal; or it can be seen as an independent mathematical quantity, a differential form. More complicated cases may vary the notation slightly. In Leibniz's notation, *dx* is interpreted an infinitesimal change in *x*. Although Leibniz's interpretation lacks rigour, his integration notation is the most common one in use today.

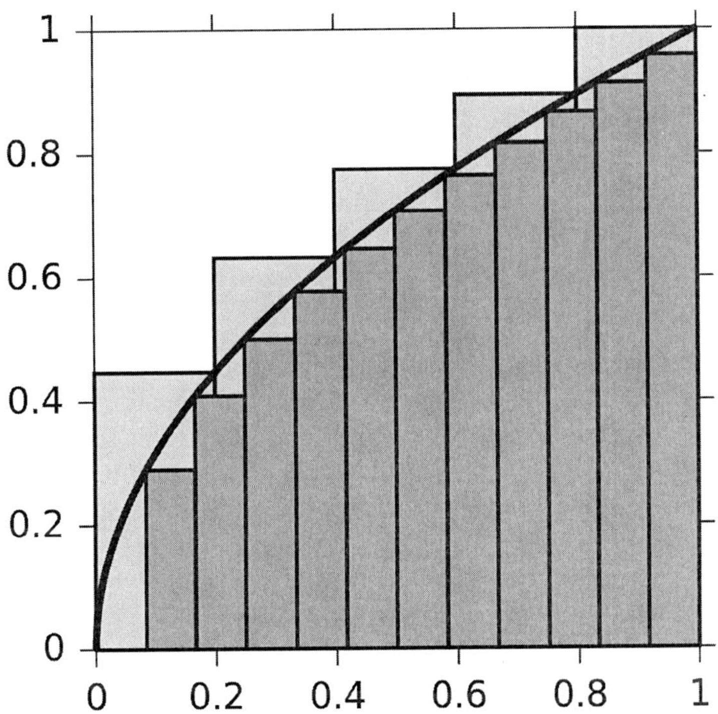

Figure: *Approximations to integral of √x from 0 to 1, with 5 right samples (above) and 12 left samples (below)*

Integrals appear in many practical situations. If a swimming pool is rectangular with a flat bottom, then from its length, width, and depth we can easily determine the volume of water it can contain (to fill it), the area of its surface (to cover it), and the length of its edge (to rope it). But if it is oval with a rounded bottom, all of these quantities call for integrals. Practical approximations may suffice for such trivial examples, but precision engineering (of any discipline) requires exact and rigorous values for these elements.

To start off, consider the curve $y = f(x)$ between $x = 0$ and $x = 1$ with $f(x) = \sqrt{x}$. We ask:

What is the area under the function f, in the interval from 0 to 1?

and call this (yet unknown) area the integral of f. The notation for this integral will be

$$\int_0^1 \sqrt{x}\,dx.$$

As a first approximation, look at the unit square given by the sides $x = 0$ to $x = 1$ and $y = f(0) = 0$ and $y = f(1) = 1$. Its area is exactly 1. As it is, the true value of the integral must be somewhat less. Decreasing the width of the approximation rectangles shall give a better result; so cross the interval in five steps, using the approximation points 0, 1/5, 2/5, and so on to 1. Fit a box for each step using the right end height of each curve piece, thus √(1D 5), √(2D 5), and so on to $\sqrt{1} = 1$. Summing the areas of these rectangles, we get a better approximation for the sought integral, namely

$$\sqrt{\frac{1}{5}}\left(\frac{1}{5}-0\right)+\sqrt{\frac{2}{5}}\left(\frac{2}{5}-\frac{1}{5}\right)+\cdots+\sqrt{\frac{5}{5}}\left(\frac{5}{5}-\frac{4}{5}\right)\approx 0.7497.$$

Notice that we are taking a sum of finitely many function values of f, multiplied with the differences of two subsequent approximation points. We can easily see that the approximation is still too large. Using more steps produces a closer approximation, but will never be exact: replacing the 5 subintervals by twelve as depicted, we will get an approximate value for the area of 0.6203, which is too small. The key idea is the transition from adding *finitely many* differences of approximation points multiplied by their respective function values to using infinitely many fine, or *infinitesimal* steps.

As for the *actual calculation of integrals*, the fundamental theorem of calculus, due to Newton and Leibniz, is the fundamental link between the operations of differentiating and integrating. Applied to the square root curve, $f(x) = x^{1/2}$, it says to look at the antiderivative $F(x) = (2/3)x^{3/2}$, and simply take $F(1) - F(0)$, where 0 and 1 are the boundaries of the interval [0,1]. So the *exact* value of the area under the curve is computed formally as

$$\int_0^1 \sqrt{x}\,dx = \int_0^1 x^{1/2}\,dx = F(1) - F(0) = 2/3.$$

(This is a case of a general rule, that for $f(x) = x^q$, with $q \neq -1$, the related function, the so-called antiderivative is $F(x) = x^{q+1}/(q + 1)$.)

The notation

$$\int f(x)dx$$

conceives the integral as a weighted sum, denoted by the elongated *s*, of function values, $f(x)$, multiplied by infinitesimal step widths, the so-called *differentials*, denoted by dx. The multiplication sign is usually omitted.

Historically, after the failure of early efforts to rigorously interpret infinitesimals, Riemann formally defined integrals as a limit of weighted sums, so that the dx suggested the limit of a difference (namely, the interval width). Shortcomings of Riemann's dependence on intervals and continuity motivated newer definitions, especially the Lebesgue integral, which is founded on an ability to extend the idea of "measure" in much more flexible ways. Thus the notation

$$\int_A f(x)d\mu$$

refers to a weighted sum in which the function values are partitioned, with μ measuring the weight to be assigned to each value. Here A denotes the region of integration.

Differential geometry, with its "calculus on manifolds", gives the familiar notation yet another interpretation. Now $f(x)$ and dx become a differential form, $\omega = f(x)\,dx$, a new differential operator d, known as the exterior derivative is introduced, and the fundamental theorem becomes the more general Stokes' theorem,

$$\int_A d\omega = \int_{\partial A} \omega,$$

from which Green's theorem, the divergence theorem, and the fundamental theorem of calculus follow.

More recently, infinitesimals have reappeared with rigor, through modern innovations such as non-standard analysis. Not only do these methods vindicate the intuitions of the pioneers; they also lead to new mathematics. Although there are differences between these conceptions of integral, there is considerable overlap. Thus, the area of the surface of the oval swimming pool can be handled as a geometric ellipse, a sum of infinitesimals, a Riemann integral, a Lebesgue integral, or as a manifold with a differential form. The calculated result will be the same for all.

Formal Definitions

There are many ways of formally defining an integral, not all of which are equivalent. The differences exist mostly to deal with differing special cases which may not be integrable under other definitions, but also occasionally for pedagogical reasons. The most commonly used definitions of integral are Riemann integrals and Lebesgue integrals.

Lebesgue Integration

In mathematics, the integral of a non-negative function can be regarded in the simplest case as the area between the graph of that function and the x-axis. Lebesgue integration is a mathematical construction that extends the integral to a larger class of functions; it also extends the domains on which these functions can be defined. It had long been understood that for non-negative functions with a smooth enough graph (such as continuous functions on closed bounded intervals) the *area under the curve* could be defined as the integral and computed using techniques of approximation of the region by polygons. However, as the need to consider more irregular functions arose (for example, as a result of the limiting processes of mathematical analysis and the mathematical theory of probability) it became clear that more careful approximation techniques would be needed to define a suitable integral. Also, we might wish to integrate on spaces more general than the real line; the Lebesgue integral provides the right abstractions needed to do this important job.

The Lebesgue integral plays an important role in the branch of mathematics called real analysis and in many other fields in the mathematical sciences, and is named after Henri Lebesgue (1875–1941) who introduced the integral in (Lebesgue 1904). It is also a pivotal portion of the axiomatic theory of probability. The term "Lebesgue integration" may refer either to the general theory of integration of a function with respect to a general measure, as introduced by Lebesgue, or to the specific case of integration of a function defined on a sub-domain of the real line with respect to Lebesgue measure.

Introduction

The integral of a function f between limits a and b can be interpreted as the area under the graph of f. This is easy to understand for familiar functions such as polynomials, but what does it mean for more exotic functions? In general, what is the class of functions for which "area under the curve" makes sense? The answer to this question has great theoretical and practical importance.

As part of a general movement toward rigour in mathematics in the nineteenth century, attempts were made to put the integral calculus on a firm foundation. The Riemann integral, proposed by Bernhard Riemann (1826–1866), is a broadly successful attempt to provide such a foundation. Riemann's definition starts with the construction of a sequence of easily-calculated areas which converge to the integral of a given function. This definition is successful in the sense that it gives the expected answer for many already-solved problems, and gives useful results for many other problems.

However, Riemann integration does not interact well with taking limits of sequences of functions, making such limiting processes difficult to analyze. This is of prime importance, for instance, in the study of Fourier series, Fourier transforms and other topics. The Lebesgue integral is better able to describe how and when it is possible to take limits under the integral sign.

The Lebesgue definition considers a different class of easily-calculated areas than the Riemann definition, which is the main reason the Lebesgue integral is better behaved. The Lebesgue definition also makes it possible to calculate integrals for a broader class of functions. For example, the Dirichlet function, which is 0 where its argument is irrational and 1 otherwise, has a Lebesgue integral, but it does not have a Riemann integral.

Lebesgue's approach to integration was summarized in a letter to Paul Montel. He writes:

I have to pay a certain sum, which I have collected in my pocket. I take the bills and coins out of my pocket and give them to the creditor in the order I find them until I have reached the total sum. This is the Riemann integral. But I can proceed differently. After I have taken all the money out of my pocket I order the bills and coins according to identical values and then I pay the several heaps one after the other to the creditor. This is my integral.

—*Source*: (Siegmund-Schultze 2008)

The insight is that one should be able to rearrange the values of a function freely while preserving the value of the integral. This process of rearrangement can convert a very pathological function into one which is "nice" from the point of view of integration, and thus allows for such pathological functions to be integrated.

Intuitive Interpretation

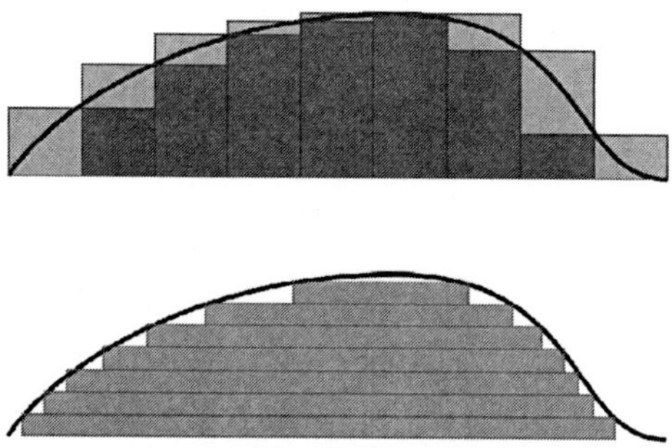

Figure: *Riemann-Darboux's integration (in blue) and Lebesgue integration (in red).*

To get some intuition about the different approaches to integration, let us imagine that it is desired to find a mountain's volume (above sea level).

The Riemann-Darboux Approach: Divide the base of the mountain into a grid of 1 metre squares. Measure the altitude of the mountain at the centre of each square. The volume on a single grid square is approximately 1×1×(altitude), so the total volume is the sum of the altitudes.

The Lebesgue Approach: Draw a contour map of the mountain, where adjacent contours are 1 metre of altitude apart. The volume of earth contained in a single contour is approximately that contour's area times 1, so the total volume is the sum of these areas times 1 metre.

Folland summarizes the difference between the Riemann and Lebesgue approaches thus: "to compute the Riemann integral of f, one partitions the domain $[a, b]$ into subintervals", while in the Lebesgue integral, "one is in effect partitioning the range of f".

Toward a Formal Definition

To define the Lebesgue integral formally requires the notion of a measure which, roughly, associates to each set A of real numbers a nonnegative number $\mu(A)$ representing the "size" of A. This notion of "size" should agree with the usual length of an interval or disjoint union of intervals. Suppose that $f : \mathrm{R} \to \mathrm{R}$ is a non-negative real-valued function. Using the "partitioning the range of f" philosophy, the integral of f should be the sum over t of the area of the thin horizontal strip between $y = t$ and $y = t+dt$. This area is just

$$\mu(\{x \mid f(x) > t\})dt.$$

Let

$$f^*(t) = \mu(\{x \mid f(x) > t\}).$$

The Lebesgue integral of f is then defined by

$$\int f = \int_0^\infty f^*(t)dt$$

where the integral on the right is an ordinary improper Riemann integral (note that f^* is a decreasing non-negative function, and therefore has a well-defined improper Riemann integral). For a suitable class of functions (the measurable functions) this defines the Lebesgue integral.

A general (not necessarily positive) function f is Lebesgue integrable if the area between the graph of f and the x-axis is finite:

$$\int |f| d\mu < +\infty.$$

In that case, the integral is, as in the Riemannian case, the difference between the area above the x-axis and the area below the x-axis:

$$\int f d\mu = \int f^+ d\mu - \int f^- d\mu$$

where

$$f^+(x) = \max(\{f(x), 0\}) = \begin{cases} f(x), & \text{if } f(x) > 0, \\ 0, & \text{otherwise} \end{cases}$$

$$f^-(x) = \max(\{-f(x), 0\}) = \begin{cases} -f(x), & \text{if } f(x) < 0, \\ 0, & \text{otherwise.} \end{cases}$$

Construction

The discussion that follows parallels the most common expository approach to the Lebesgue integral. In this approach, the theory of integration has two distinct parts:

1. A theory of measurable sets and measures on these sets.
2. A theory of measurable functions and integrals on these functions.

The function whose integral is to be found is then approximated by certain so-called simple functions, whose integrals can be written in terms of the measure. The integral of the original function is then the limit of the integral of the simple functions.

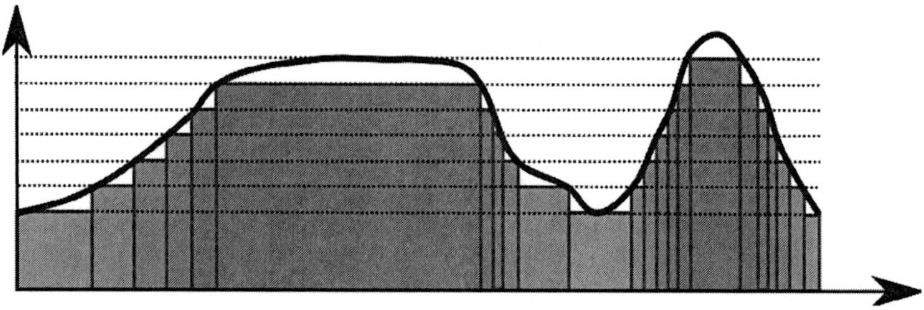

Figure: *Approximating a function by simple functions.*

Measure Theory

Measure theory was initially created to provide a useful abstraction of the notion of length of subsets of the real line and, more generally, area and volume of subsets of Euclidean spaces. In particular, it provided a systematic answer to the question of which subsets of R have a length. As was shown by later developments in set theory, it is actually impossible to assign a length to all subsets of R in a way which preserves some natural additivity and translation invariance properties. This suggests that picking out a suitable class of *measurable* subsets is an essential prerequisite.

The Riemann integral uses the notion of length explicitly. Indeed, the element of calculation for the Riemann integral is the rectangle $[a, b] \times [c, d]$, whose area is calculated to be $(b - a)(d - c)$. The quantity $b - a$ is the length of the base of the rectangle and $d - c$ is the height of the rectangle. Riemann could only use planar rectangles to approximate the area under the curve because there was no adequate theory for measuring more general sets.

In the development of the theory in most modern textbooks (after 1950), the approach to measure and integration is *axiomatic*. This means that a measure is any function μ defined on a certain class X of subsets of a set E, which satisfies a certain list of properties. These properties can be shown to hold in many different cases.

Integration

We start with a measure space (E, X, μ) where E is a set, X is a σ-algebra of subsets of E and μ is a (non-negative) measure on E, defined on the sets of X.

For example, E can be Euclidean n-space R^n or some Lebesgue measurable subset of it, X will be the σ-algebra of all Lebesgue measurable subsets of E, and μ will be the Lebesgue measure. In the mathematical theory of probability, we confine our study to a probability measure μ, which satisfies $\mu(E) = 1$.

In Lebesgue's theory, integrals are defined for a class of functions called measurable functions. A function f is measurable if the pre-image of every interval of the form (t, ∞) is in X:

$$\{x|f(x) > t\} \in X \quad \text{for all } t \in \mathrm{R}.$$

It can be shown that this is equivalent to requiring that the pre-image of any Borel subset of R be in X. We will make this assumption henceforth. The set of measurable functions is closed under algebraic operations, but more importantly it is closed under various kinds of pointwise sequential limits:

$$\sup_{k \in \mathrm{N}} f_k, \quad \liminf_{k \in \mathrm{N}} f_k, \quad \limsup_{k \in \mathrm{N}} f_k$$

are measurable if the original sequence $(f_k)_k$, where $k \in$ N, consists of measurable functions.

We build up an integral

$$\int_E f \, d\mu = \int_E f(x) \mu(dx)$$

for measurable real-valued functions f defined on E in stages:

Indicator functions: To assign a value to the integral of the indicator function 1_S of a measurable set S consistent with the given measure μ, the only reasonable choice is to set:

$$\int 1_S \, d\mu = \mu(S).$$

Notice that the result may be equal to $+\infty$, unless μ is a *finite* measure.

Simple functions: A finite linear combination of indicator functions

$$\sum_k a_k 1_{S_k}$$

where the coefficients a_k are real numbers and the sets S_k are measurable, is called a measurable simple function. We extend the integral by linearity to *non-negative* measurable simple functions. When the coefficients a_k are non-negative, we set

$$\int \left(\sum_k a_k 1_{S_k} \right) d\mu = \sum_k a_k \int 1_{S_k} \, d\mu = \sum_k a_k \, \mu(S_k).$$

The convention $0 \times \infty = 0$ must be used, and the result may be infinite. Even if a simple function can be written in many ways as a linear combination of indicator functions, the integral will always be the same; this can be shown using the additivity property of measures.

Some care is needed when defining the integral of a *real-valued* simple function, to avoid the undefined expression $\infty - \infty$: one assumes that the representation

$$f = \sum_k a_k 1_{S_k}$$

is such that $\mu(S_k) < \infty$ whenever $a_k \neq 0$. Then the above formula for the integral of f makes sense, and the result does not depend upon the particular representation of f satisfying the assumptions.

If B is a measurable subset of E and s is a measurable simple function one defines

$$\int_B s d\mu = \int 1_B s d\mu = \sum_k a_k \mu(S_k \cap B).$$

Non-negative Functions: Let f be a non-negative measurable function on E which we allow to attain the value +", in other words, f takes non-negative values in the extended real number line. We define

$$\int_E f d\mu = \sup\left\{ \int_E s d\mu : 0 \leq s \leq f, s \text{ simple} \right\}.$$

We need to show this integral coincides with the preceding one, defined on the set of simple functions. When E is a segment $[a, b]$, there is also the question of whether this corresponds in any way to a Riemann notion of integration. It is possible to prove that the answer to both questions is yes.

We have defined the integral of f for any non-negative extended real-valued measurable function on E. For some functions, this integral $\int_E f \, d\mu$ will be infinite.

Signed Functions: To handle signed functions, we need a few more definitions. If f is a measurable function of the set E to the reals (including $\pm\infty$), then we can write

$$f = f^+ - f^-,$$

where

$$f^+(x) = \begin{cases} f(x) & \text{if } f(x) > 0 \\ 0 & \text{otherwise} \end{cases}$$

$$f^{-}(x) = \begin{cases} -f(x) & \text{if } f(x) < 0 \\ 0 & \text{otherwise} \end{cases}$$

Note that both f^+ and f^- are non-negative measurable functions. Also note that

$$|f| = f^{+} + f^{-}.$$

We say that the Lebesgue integral of the measurable function f *exists*, or *is defined* if at least one of $\int f^{+}\,d\mu$ and $\int f^{-}\,d\mu$ is finite:

$$\min\left(\int f^{+}\,d\mu, \int f^{-}\,d\mu\right) < \infty.$$

In this case we *define*

$$\int f\,d\mu = \int f^{+}\,d\mu - \int f^{-}\,d\mu.$$

If

$$\int |f|\,d\mu < \infty,$$

we say that f is *Lebesgue integrable.*

It turns out that this definition gives the desirable properties of the integral.

Complex valued functions can be similarly integrated, by considering the real part and the imaginary part separately.

Example

Consider the indicator function of the rational numbers, 1_Q. This function is nowhere continuous.

- 1_Q is not Riemann-integrable on $[0,1]$: No matter how the set $[0,1]$ is partitioned into subintervals, each partition will contain at least one rational and at least one irrational number, because rationals and irrationals are both dense in the reals. Thus the upper Darboux sums will all be one, and the lower Darboux sums will all be zero.
- 1_Q is Lebesgue-integrable on $[0,1]$ using the Lebesgue measure: Indeed it is the indicator function of the rationals so by definition

$$\int_{[0,1]} 1_Q\,d\mu = \mu(Q \cap [0,1]) = 0,$$

because Q is countable.

Domain of Integration

A technical issue in Lebesgue integration is that the domain of integration is defined as a *set* (a subset of a measure space), with no notion of orientation. In elementary calculus, one defines integration with respect to an orientation:

$$\int_b^a f := -\int_a^b f.$$

Generalizing this to higher dimensions yields integration of differential forms. By contrast, Lebesgue integration provides an alternative generalization, integrating over subsets with respect to a measure; this can be notated as

$$\int_A f d\mu = \int_{[a,b]} f d\mu$$

to indicate integration over a subset A.

Limitations of the Riemann Integral

Here we discuss the limitations of the Riemann integral and the greater scope offered by the Lebesgue integral. We presume a working understanding of the Riemann integral. With the advent of Fourier series, many analytical problems involving integrals came up whose satisfactory solution required interchanging limit processes and integral signs. However, the conditions under which the integrals

$$\sum_k \int f_k(x)dx, \quad \int\left[\sum_k f_k(x)\right]dx$$

are equal proved quite elusive in the Riemann framework. There are some other technical difficulties with the Riemann integral. These are linked with the limit-taking difficulty discussed above.

Failure of monotone convergence. As shown above, the indicator function 1_Q on the rationals is not Riemann integrable. In particular, the Monotone convergence theorem fails. To see why, let $\{a_k\}$ be an enumeration of all the rational numbers in [0,1] (they are countable so this can be done.) Then let

$$g_k(x) = \begin{cases} 1 & \text{if } x = a_j, j \leq k \\ 0 & \text{otherwise} \end{cases}$$

The function g_k is zero everywhere except on a finite set of points, hence its Riemann integral is zero. The sequence g_k is also clearly non-negative and monotonically increasing to 1_Q, which is not Riemann integrable.

Unsuitability for unbounded intervals. The Riemann integral can only integrate functions on a bounded interval. It can however be extended to unbounded intervals by taking limits, so long as this doesn't yield an answer such as $\infty - \infty$.

Integrating on structures other than Euclidean space. The Riemann integral is inextricably linked to the order structure of the line.

Basic Theorems of the Lebesgue Integral

The Lebesgue integral does not distinguish between functions which differ only on a set of μ-measure zero. To make this precise, functions f and g are said to be equal almost everywhere (a.e.) if

$$\mu(\{x \in E : f(x) \neq g(x)\}) = 0.$$

- If f, g are non-negative measurable functions (possibly assuming the value $+\infty$) such that $f = g$ almost everywhere, then

$$\int f \, d\mu = \int g \, d\mu.$$

To wit, the integral respects the equivalence relation of almost-everywhere equality.

- If f, g are functions such that $f = g$ almost everywhere, then f is Lebesgue integrable if and only if g is Lebesgue integrable, and the integrals of f and g are the same if they exist.

The Lebesgue integral has the following properties:

Linearity: If f and g are Lebesgue integrable functions and a and b are real numbers, then $af + bg$ is Lebesgue integrable and

$$\int (af + bg) d\mu = a \int f \, d\mu + b \int g \, d\mu.$$

Monotonicity: If $f \leq g$, then

$$\int f \, d\mu \leq \int g \, d\mu.$$

Monotone convergence theorem: Suppose $\{f_k\}_{k \in N}$ is a sequence of non-negative measurable functions such that

$$f_k(x) \leq f_{k+1}(x) \quad \forall k \in N, \forall x \in E.$$

Then, the pointwise limit f of f_k is Lebesgue integrable and

$$\lim_k \int f_k \, d\mu = \int f \, d\mu.$$

Note: The value of any of the integrals is allowed to be infinite.

Fatou's lemma: If $\{f_k\}_{k \in N}$ is a sequence of non-negative measurable functions, then $\int \liminf_k f_k \, d\mu \leq \liminf_k \int f_k \, d\mu$.

Again, the value of any of the integrals may be infinite.

Dominated convergence theorem: Suppose $\{f_k\}_{k \in N}$ is a sequence of complex measurable functions with pointwise limit f, and there is a Lebesgue integrable function g (i.e., g belongs to the space L^1) such that $|f_k| \leq g$ for all k.

Then, f is Lebesgue integrable and

$$\lim_k \int f_k \, d\mu = \int f \, d\mu.$$

Proof Techniques

To illustrate some of the proof techniques used in Lebesgue integration theory, we sketch a proof of the above mentioned Lebesgue monotone convergence theorem. Let $\{f_k\}_{k \in N}$ be a non-decreasing sequence of non-negative measurable functions and put

$$f = \sup_{k \in N} f_k = \lim_{k \in N} f_k.$$

By the monotonicity property of the integral, it is immediate that:

$$\int f \, d\mu \geq \lim_k \int f_k \, d\mu$$

and the limit on the right exists, because the sequence is monotonic. We now prove the inequality in the other direction. It follows from the definition of integral that there is a non-decreasing sequence (g_n) of non-negative simple functions such that $g_n \leq f$ and

$$\lim_n \int g_n \, d\mu = \int f \, d\mu.$$

Therefore, it suffices to prove that for each $n \in N$,

$$\int g_n \, d\mu \leq \lim_k \int f_k \, d\mu.$$

We will show that if g is a simple function and

$$\lim_k f_k(x) \geq g(x)$$

almost everywhere, then

$$\lim_k \int f_k \, d\mu \geq \int g \, d\mu.$$

By breaking up the function g into its constant value parts, this reduces to the case in which g is the indicator function of a set. The result we have to prove is then

Suppose A is a measurable set and $\{f_k\}_{k \in N}$ is a nondecreasing sequence of non-negative measurable functions on E such that

$$\lim_k f_k(x) \geq 1$$

for almost all $x \in A$. Then

$$\lim_k \int f_k \, d\mu \geq \mu(A).$$

To prove this result, fix $\varepsilon > 0$ and define the sequence of measurable sets

$$B_k = \{x \in A : f_k(x) \geq 1 - \varepsilon\}.$$

By monotonicity of the integral, it follows that for any $k \in N$,

$$(1-\varepsilon)\mu(B_k) = \int (1-\varepsilon)1_{B_k} \, d\mu \leq \int f_k \, d\mu$$

Because almost every x will be in B_k for large enough k, we have

$$\bigcup_k B_k = A,$$

up to a set of measure 0. Thus by countable additivity of μ, and because B_k increases with k,

$$\mu(A) = \lim_k \mu(B_k) \leq \lim_k (1-\varepsilon)^{-1} \int f_k \, d\mu.$$

As this is true for any positive ε the result follows.

Alternative Formulations

It is possible to develop the integral with respect to the Lebesgue measure without relying on the full machinery of measure theory. One such approach is provided by the Daniell integral.

There is also an alternative approach to developing the theory of integration via methods of functional analysis. The Riemann integral exists for any continuous function f of compact support defined on R^n (or a fixed open subset). Integrals of more general functions can be built starting from these integrals.

Let C_c be the space of all real-valued compactly supported continuous functions of R. Define a norm on C_c by

$$\| f \| = \int |f(x)| \, dx.$$

Then C_c is a normed vector space (and in particular, it is a metric space.) All metric spaces have Hausdorff completions, so let L^1 be its completion. This space is isomorphic to the space of Lebesgue integrable functions modulo the subspace of functions with integral zero.

Furthermore, the Riemann integral $\int$ is a uniformly continuous functional with respect to the norm on C_c, which is dense in L^1. Hence $\int$ has a unique extension to all of L^1. This integral is precisely the Lebesgue integral.

More generally, when the measure space on which the functions are defined is also a locally compact topological space (as is the case with the real numbers R), measures compatible with the topology in a suitable sense (Radon measures, of which the Lebesgue measure is an example) an integral with respect to them can be defined in the same manner, starting from the integrals of continuous functions with compact support. More precisely, the compactly supported functions form a vector space that carries a natural topology, and a (Radon) measure is defined as a continuous linear functional on this space. The value of a measure at a compactly supported function is then also by definition the integral of the function. One then proceeds to expand the measure (the integral) to more general functions by continuity, and defines the measure of a set as the integral of its indicator function. This is the approach taken by Bourbaki (2004) and a certain number of other authors.

Limitations of Lebesgue Integral

The main purpose of Lebesgue integral is to provide an integral notation where limits of integrals hold under mild assumptions. There is no guarantee that every function is Lebesgue integrable. It may happen that improper (Riemann) integrals exist for functions that are not Lebesgue integrable. One example would be

$$\frac{\sin(x)}{x}.$$

This function is not Lebesgue integrable as

$$\int\left|\frac{\sin(x)}{x}\right|d\mu = \infty.$$

On the other hand, it exists as an improper Riemann integral and the integral can be computed to be finite.

Other Integrals

Although the Riemann and Lebesgue integrals are the most widely used definitions of the integral, a number of others exist, including:

- The Darboux integral which is equivalent to a Riemann integral, meaning that a function is Darboux-integrable if and only if it is Riemann-integrable, and the values of the two integrals, if they exist, are equal. Darboux integrals have the advantage of being simpler to define than Riemann integrals.

- The Riemann–Stieltjes integral, an extension of the Riemann integral.
- The Lebesgue-Stieltjes integral, further developed by Johann Radon, which generalizes the Riemann–Stieltjes and Lebesgue integrals.
- The Daniell integral, which subsumes the Lebesgue integral and Lebesgue-Stieltjes integral without the dependence on measures.
- The Haar integral, used for integration on locally compact topological groups, introduced by Alfréd Haar in 1933.
- The Henstock–Kurzweil integral, variously defined by Arnaud Denjoy, Oskar Perron, and (most elegantly, as the gauge integral) Jaroslav Kurzweil, and developed by Ralph Henstock.
- The Itô integral and Stratonovich integral, which define integration with respect to semimartingales such as Brownian motion.
- The Young integral, which is a kind of Riemann–Stieltjes integral with respect to certain functions of unbounded variation.
- The rough path integral defined for functions equipped with some additional "rough path" structure, generalizing stochastic integration against both semimartingales and processes such as the fractional Brownian motion.

Properties

Linearity:

- The collection of Riemann integrable functions on a closed interval $[a, b]$ forms a vector space under the operations of pointwise addition and multiplication by a scalar, and the operation of integration

$$f \mapsto \int_a^b f\, dx$$

is a linear functional on this vector space. Thus, firstly, the collection of integrable functions is closed under taking linear combinations; and, secondly, the integral of a linear combination is the linear combination of the integrals,

$$\int_a^b (\alpha f + \beta g)(x)dx = \alpha \int_a^b f(x)dx + \beta \int_a^b g(x)dx.$$

- Similarly, the set of real-valued Lebesgue integrable functions on a given measure space E with measure μ is closed under taking linear combinations and hence form a vector space, and the Lebesgue integral

$$f \mapsto \int_E f d\mu$$

is a linear functional on this vector space, so that

$$\int_E (\alpha f + \beta g) d\mu = \alpha \int_E f d\mu + \beta \int_E g d\mu.$$

- More generally, consider the vector space of all measurable functions on a measure space (E,μ), taking values in a locally compact complete topological vector space V over a locally compact topological field K, $f : E \to V$. Then one may define an abstract integration map assigning to each function f an element of V or the symbol ∞,

$$f \mapsto \int_E f d\mu,$$

that is compatible with linear combinations. In this situation the linearity holds for the subspace of functions whose integral is an element of V (i.e. "finite"). The most important special cases arise when K is R, C, or a finite extension of the field Q_p of p-adic numbers, and V is a finite-dimensional vector space over K, and when K=C and V is a complex Hilbert space.

Linearity, together with some natural continuity properties and normalisation for a certain class of "simple" functions, may be used to give an alternative definition of the integral. This is the approach of Daniell for the case of real-valued functions on a set X, generalized by Nicolas Bourbaki to functions with values in a locally compact topological vector space.

Inequalities for Integrals

A number of general inequalities hold for Riemann-integrable functions defined on a closed and bounded interval $[a, b]$ and can be generalized to other notions of integral (Lebesgue and Daniell).

- *Upper and lower bounds.* An integrable function f on $[a, b]$, is necessarily bounded on that interval. Thus there are real numbers m and M so that $m \le f(x) \le M$ for all x in $[a, b]$. Since the lower and upper sums of f over $[a, b]$ are therefore bounded by, respectively, $m(b - a)$ and $M(b - a)$, it follows that

$$m(b-a) \le \int_a^b f(x)dx \le M(b-a).$$

- *Inequalities between functions.* If $f(x) \le g(x)$ for each x in $[a, b]$ then each of the upper and lower sums of f is bounded above by the upper and lower sums, respectively, of g. Thus

$$\int_a^b f(x)dx \le \int_a^b g(x)dx.$$

This is a generalization of the above inequalities, as $M(b - a)$ is the integral of the constant function with value M over $[a, b]$.

In addition, if the inequality between functions is strict, then the inequality between integrals is also strict. That is, if $f(x) < g(x)$ for each x in $[a, b]$, then

$$\int_a^b f(x)dx < \int_a^b g(x)dx.$$

- *Subintervals.* If $[c, d]$ is a subinterval of $[a, b]$ and $f(x)$ is non-negative for all x, then

$$\int_c^d f(x)dx \le \int_a^b f(x)dx.$$

- *Products and absolute values of functions.* If f and g are two functions then we may consider their pointwise products and powers, and absolute values:

$$(fg)(x) = f(x)g(x),\ f^2(x) = (f(x))^2,\ |f|(x) = |f(x)|.$$

If f is Riemann-integrable on $[a, b]$ then the same is true for $|f|$, and

$$\left|\int_a^b f(x)dx\right| \le \int_a^b |f(x)|dx.$$

Moreover, if f and g are both Riemann-integrable then f^2, g^2, and fg are also Riemann-integrable, and

$$\left(\int_a^b (fg)(x)dx\right)^2 \le \left(\int_a^b f(x)^2 dx\right)\left(\int_a^b g(x)^2 dx\right).$$

This inequality, known as the Cauchy–Schwarz inequality, plays a prominent role in Hilbert space theory, where the left hand side is interpreted as the inner product of two square-integrable functions f and g on the interval $[a, b]$.

- *Hölder's inequality.* Suppose that p and q are two real numbers, $1 \le p, q \le \infty$ with $1/p + 1/q = 1$, and f and g are two Riemann-integrable functions. Then the functions $|f|^p$ and $|g|^q$ are also integrable and the following Hölder's inequality holds:

$$\left|\int f(x)g(x)dx\right| \leq \left(\int |f(x)|^p dx\right)^{1/p} \left(\int |g(x)|^q dx\right)^{1/q}.$$

For $p = q = 2$, Hölder's inequality becomes the Cauchy–Schwarz inequality.

- *Minkowski inequality.* Suppose that $p \geq 1$ is a real number and f and g are Riemann-integrable functions. Then $|f|^p$, $|g|^p$ and $|f + g|^p$ are also Riemann integrable and the following Minkowski inequality holds:

$$\left(\int |f(x)+g(x)|^p dx\right)^{1/p} \leq \left(\int |f(x)|^p dx\right)^{1/p} + \left(\int |g(x)|^p dx\right)^{1/p}.$$

An analogue of this inequality for Lebesgue integral is used in construction of L^p spaces.

Conventions

In this section f is a real-valued Riemann-integrable function. The integral

$$\int_a^b f(x)dx$$

over an interval $[a, b]$ is defined if $a < b$. This means that the upper and lower sums of the function f are evaluated on a partition $a = x_0 \leq x_1 \leq \ldots \leq x_n = b$ whose values x_i are increasing. Geometrically, this signifies that integration takes place "left to right", evaluating f within intervals $[x\ , x_{i+1}]$ where an interval with a higher index lies to the right of one with a lower index. The values a and b, the end-points of the interval, are called the limits of integration of f. Integrals can also be defined if $a > b$:

- *Reversing limits of integration.* If $a > b$ then define

$$\int_a^b f(x)dx = -\int_b^a f(x)dx.$$

This, with $a = b$, implies:

- *Integrals over intervals of length zero.* If a is a real number then

$$\int_a^a f(x)dx = 0.$$

The first convention is necessary in consideration of taking integrals over subintervals of $[a, b]$; the second says that an integral taken over a degenerate interval, or a point, should be zero. One reason for the first convention is that the integrability of f on an interval $[a, b]$ implies that f is integrable on any subinterval $[c, d]$, but in particular integrals have the property that:

- *Additivity of integration on intervals.* If c is any element of $[a, b]$, then

$$\int_a^b f(x)dx = \int_a^c f(x)dx + \int_c^b f(x)dx.$$

With the first convention the resulting relation

$$\begin{aligned}\int_a^c f(x)dx &= \int_a^b f(x)dx - \int_c^b f(x)dx \\ &= \int_a^b f(x)dx + \int_b^c f(x)dx\end{aligned}$$

is then well-defined for any cyclic permutation of a, b, and c.

Instead of viewing the above as conventions, one can also adopt the point of view that integration is performed of differential forms on *oriented* manifolds only. If M is such an oriented m-dimensional manifold, and M' is the same manifold with opposed orientation and ω is an m-form, then one has:

$$\int_M \omega = -\int_{M'} \omega.$$

These conventions correspond to interpreting the integrand as a differential form, integrated over a chain. In measure theory, by contrast, one interprets the integrand as a function f with respect to a measure μ, and integrates over a subset A, without any notion of orientation; one writes $\int_A f\,d\mu = \int_{[a,b]} f\,d\mu$ to indicate integration over a subset A. This is a minor distinction in one dimension, but becomes subtler on higher dimensional manifolds.

Fundamental Theorem of Calculus

The *fundamental theorem of calculus* is the statement that differentiation and integration are inverse operations: if a continuous function is first integrated and then differentiated, the original function is retrieved. An important consequence, sometimes called the *second fundamental theorem of calculus*, allows one to compute integrals by using an antiderivative of the function to be integrated.

Statements of Theorems

- *Fundamental theorem of calculus.* Let f be a continuous real-valued function defined on a closed interval $[a, b]$. Let F be the function defined, for all x in $[a, b]$, by

$$F(x) = \int_a^x f(t)dt.$$

Then, F is continuous on $[a, b]$, differentiable on the open interval (a, b), and

$$F'(x) = f(x)$$

for all x in (a, b).

- *Second fundamental theorem of calculus.* Let f be a real-valued function defined on a closed interval $[a, b]$ that admits an antiderivative g on $[a, b]$. That is, f and g are functions such that for all x in $[a, b]$,

$$f(x) = g'(x).$$

If f is integrable on $[a, b]$ then

$$\int_a^b f(x)dx = g(b) - g(a).$$

Extensions

Improper integrals:

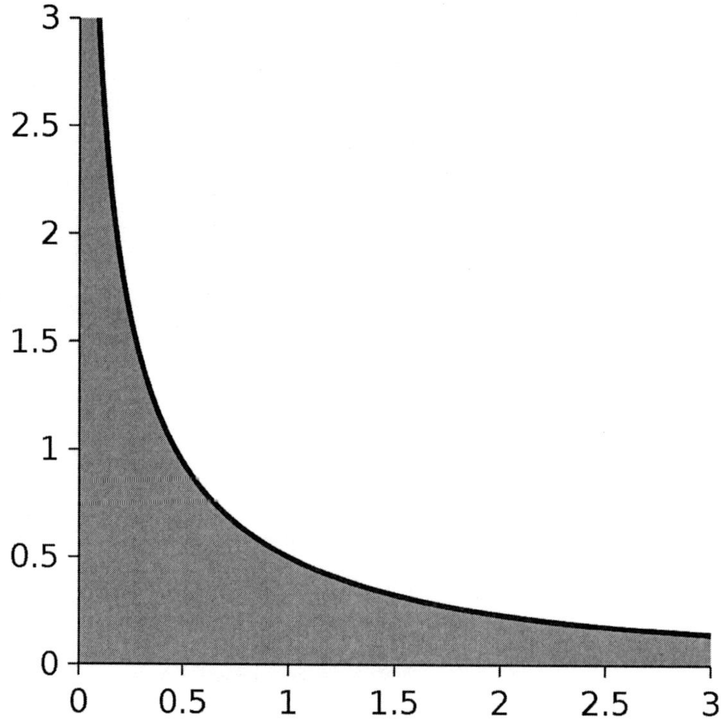

Figure: *The improper integral* $\int_0^\infty \frac{dx}{(x+1)\sqrt{x}} = \pi$ *has unbounded intervals for both domain and range.*

A "proper" Riemann integral assumes the integrand is defined and finite on a closed and bounded interval, bracketed by the limits of integration. An improper integral occurs when one or more of these conditions is not satisfied. In some cases such integrals may be defined by considering the limit of a sequence of proper Riemann integrals on progressively larger intervals.

If the interval is unbounded, for instance at its upper end, then the improper integral is the limit as that endpoint goes to infinity.

$$\int_a^\infty f(x)dx = \lim_{b\to\infty} \int_a^b f(x)dx$$

If the integrand is only defined or finite on a half-open interval, for instance $(a,b]$, then again a limit may provide a finite result.

$$\int_a^b f(x)dx = \lim_{\epsilon\to 0} \int_{a+\epsilon}^b f(x)dx$$

That is, the improper integral is the limit of proper integrals as one endpoint of the interval of integration approaches either a specified real number, or ∞, or $-\infty$. In more complicated cases, limits are required at both endpoints, or at interior points.

Consider, for example, the function $1/((x+1)\sqrt{x})$ integrated from 0 to ∞ (shown right). At the lower bound, as x goes to 0 the function goes to ∞, and the upper bound is itself ∞, though the function goes to 0. Thus this is a doubly improper integral. Integrated, say, from 1 to 3, an ordinary Riemann sum suffices to produce a result of $\pi/6$. To integrate from 1 to ∞, a Riemann sum is not possible. However, any finite upper bound, say t (with $t > 1$), gives a well-defined result, $2\arctan(\sqrt{t}) - \pi/2$. This has a finite limit as t goes to infinity, namely $\pi/2$. Similarly, the integral from 1/3 to 1 allows a Riemann sum as well, coincidentally again producing $\pi/6$. Replacing 1/3 by an arbitrary positive value s (with $s < 1$) is equally safe, giving $\pi/2 - 2\arctan(\sqrt{s})$. This, too, has a finite limit as s goes to zero, namely $\pi/2$. Combining the limits of the two fragments, the result of this improper integral is

$$\begin{aligned}
\int_0^\infty \frac{dx}{(x+1)\sqrt{x}} &= \lim_{s\to 0}\int_s^1 \frac{dx}{(x+1)\sqrt{x}} + \lim_{t\to\infty}\int_1^t \frac{dx}{(x+1)\sqrt{x}} \\
&= \lim_{s\to 0}\left(\frac{\pi}{2} - 2\arctan\sqrt{s}\right) + \lim_{t\to\infty}\left(2\arctan\sqrt{t} - \frac{\pi}{2}\right) \\
&= \frac{\pi}{2} + \left(\pi - \frac{\pi}{2}\right) \\
&= \frac{\pi}{2} + \frac{\pi}{2} \\
&= \pi.
\end{aligned}$$

This process does not guarantee success; a limit may fail to exist, or may be unbounded. For example, over the bounded interval 0 to 1 the integral of $1/x$ does not converge; and over the unbounded interval 1 to ∞ the integral of $1/\sqrt{x}$ does not converge.

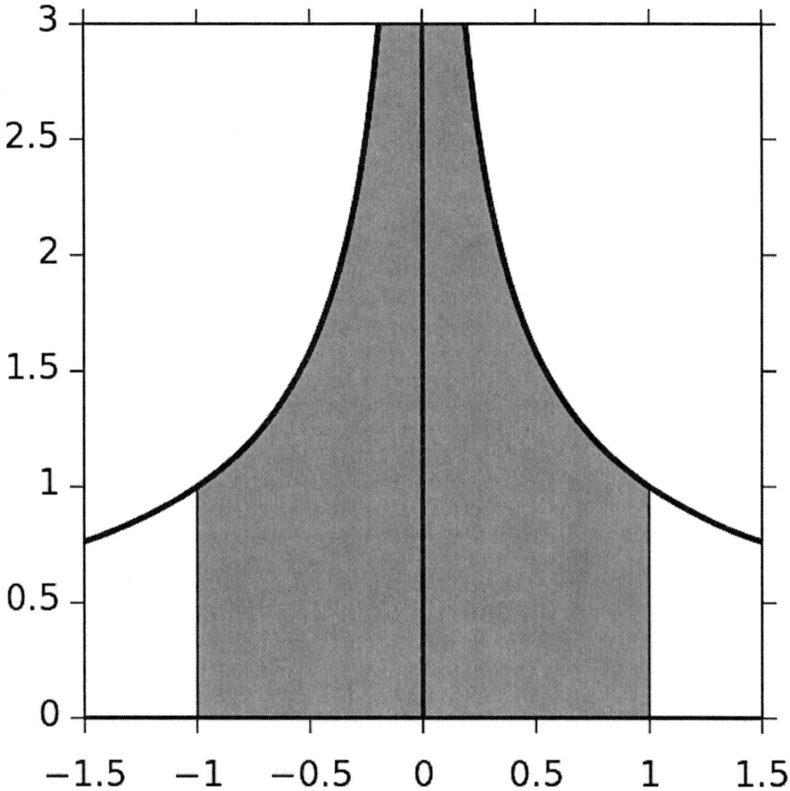

***Figure:** The improper integral* $\int_{-1}^{1} \frac{dx}{\sqrt[3]{x^2}} = 6$ *is unbounded internally, but both left and right limits exist.*

It may also happen that an integrand is unbounded at an interior point, in which case the integral must be split at that point, and the limit integrals on both sides must exist and must be bounded. Thus

$$\begin{aligned}\int_{-1}^{1} \frac{dx}{\sqrt[3]{x^2}} &= \lim_{s\to 0} \int_{-1}^{-s} \frac{dx}{\sqrt[3]{x^2}} + \lim_{t\to 0} \int_{t}^{1} \frac{dx}{\sqrt[3]{x^2}} \\ &= \lim_{s\to 0} 3(1-\sqrt[3]{s}) + \lim_{t\to 0} 3(1-\sqrt[3]{t}) \\ &= 3+3 \\ &= 6.\end{aligned}$$

But the similar integral

$$\int_{-1}^{1} \frac{dx}{x}$$

cannot be assigned a value in this way, as the integrals above and below zero do not independently converge.

Multiple Integration

Figure: *Double integral as volume under a surface.*

Integrals can be taken over regions other than intervals. In general, an integral over a set E of a function f is written:

$$\int_E f(x)dx.$$

Here x need not be a real number, but can be another suitable quantity, for instance, a vector in R^3. Fubini's theorem shows that such integrals can be rewritten as an *iterated integral*. In other words, the integral can be calculated by integrating one coordinate at a time.

Just as the definite integral of a positive function of one variable represents the area of the region between the graph of the function and the x-axis, the *double integral* of a positive function of two variables represents the volume of the region between the surface defined by the function and the plane which contains its domain. (The same volume can be obtained via the *triple integral* — the integral of a function in three variables — of the constant function $f(x, y, z) = 1$ over the above mentioned region between the surface and the plane.) If the number of variables is higher, then the integral represents a hypervolume, a volume of a solid of more than three dimensions that cannot be graphed.

For example, the volume of the cuboid of sides $4 \times 6 \times 5$ may be obtained in two ways:

- By the double integral

$$\iint_D 5\,dx\,dy$$

of the function $f(x, y) = 5$ calculated in the region D in the xy-plane which is the base of the cuboid. For example, if a rectangular base of such a cuboid is given via the xy inequalities $3 \le x \le 7$, $4 \le y \le 10$, our above double integral now reads

$$\int_4^{10}\left[\int_3^7 5\,dx\right]dy$$

From here, integration is conducted with respect to either x or y first; in this example, integration is first done with respect to x as the interval corresponding to x is the inner integral. Once the first integration is completed via the $F(b) - F(a)$ method or otherwise, the result is again integrated with respect to the other variable. The result will equate to the volume under the surface.

- By the triple integral

$$\iiint_{\text{cuboid}} 1\,dx\,dy\,dz$$

of the constant function 1 calculated on the cuboid itself.

Line Integrals

The concept of an integral can be extended to more general domains of integration, such as curved lines and surfaces. Such integrals are known as line integrals and surface integrals respectively. These have important applications in physics, as when dealing with vector fields.

A *line integral* (sometimes called a *path integral*) is an integral where the function to be integrated is evaluated along a curve. Various different line integrals are in use. In the case of a closed curve it is also called a *contour integral.*

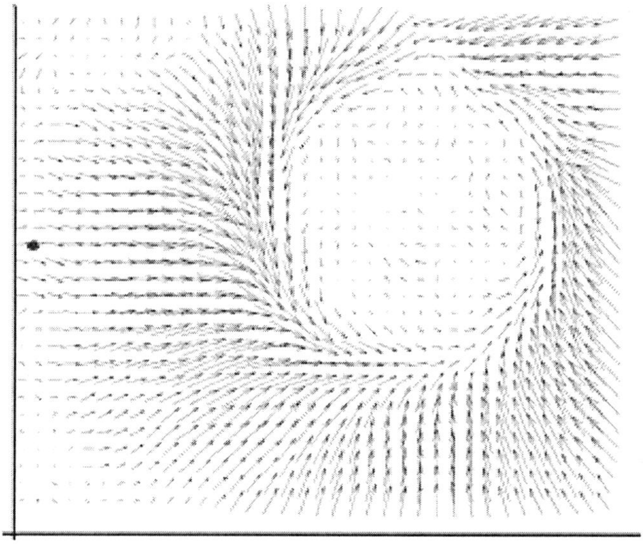

Figure: *A line integral sums together elements along a curve.*

The function to be integrated may be a scalar field or a vector field. The value of the line integral is the sum of values of the field at all points on the curve, weighted by some scalar function on the curve (commonly arc length or, for a vector field, the scalar product of the vector field with a differential vector in the curve). This weighting distinguishes the line integral from simpler integrals defined on intervals. Many simple formulas in physics have natural continuous analogs in terms of line integrals; for example, the fact that work is equal to force, F, multiplied by displacement, s, may be expressed (in terms of vector quantities) as:

$$W = \vec{F} \cdot \vec{s}.$$

For an object moving along a path in a vector field $\vec{F}$ such as an electric field or gravitational field, the total work done by the field on the object is obtained by summing up the differential work done in moving from $\vec{s}$ to $\vec{s} + d\vec{s}$. This gives the line integral

$$W = \int_C \vec{F} \cdot d\vec{s}.$$

Surface Integrals

A *surface integral* is a definite integral taken over a surface (which may be a curved set in space); it can be thought of as the double integral analog of the line integral. The function to be integrated may be a scalar field or a vector field. The value of the surface integral is the sum of the field at all points on the surface. This can be achieved by splitting the surface into surface elements, which provide the partitioning for Riemann sums.

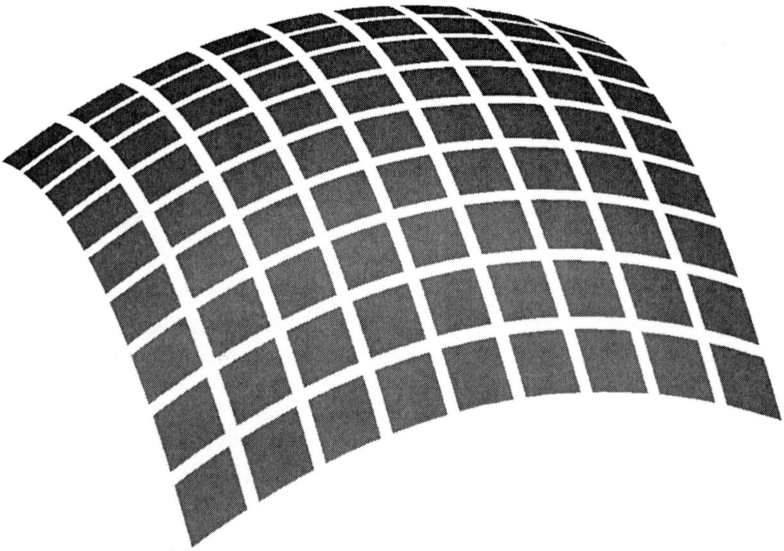

Figure: *The definition of surface integral relies on splitting the surface into small surface elements.*

For an example of applications of surface integrals, consider a vector field v on a surface S; that is, for each point x in S, $v(x)$ is a vector. Imagine that we have a fluid flowing through S, such that v(x) determines the velocity of the fluid at x. The flux is defined as the quantity of fluid flowing through S in unit amount of time. To find the flux, we need to take the dot product of v with the unit surface normal to S at each point, which will give us a scalar field, which we integrate over the surface:

$$\int_S \mathrm{v}\cdot d\mathrm{S}.$$

The fluid flux in this example may be from a physical fluid such as water or air, or from electrical or magnetic flux. Thus surface integrals have applications in physics, particularly with the classical theory of electromagnetism.

Integrals of Differential Forms

A differential form is a mathematical concept in the fields of multivariable calculus, differential topology and tensors. The modern notation for the differential form, as well as the idea of the differential forms as being the wedge products of exterior derivatives forming an exterior algebra, was introduced by Élie Cartan. We initially work in an open set in R^n. A 0-form is defined to be a smooth function f. When we integrate a function f over an m-dimensional subspace S of R^n, we write it as

$$\int_S f\,dx^1 \cdots dx^m.$$

(The superscripts are indices, not exponents.) We can consider dx^1 through dx^n to be formal objects themselves, rather than tags appended to make integrals look like Riemann sums. Alternatively, we can view them as covectors, and thus a measure of "density" (hence integrable in a general sense). We call the $dx^1, \ldots, dx^n$ *basic 1-forms.*

We define the wedge product, "$\wedge$", a bilinear "multiplication" operator on these elements, with the *alternating* property that

$$dx^a \wedge dx^a = 0$$

for all indices a. Note that alternation along with linearity and associativity implies $dx^b \wedge dx^a = -dx^a \wedge dx^b$. This also ensures that the result of the wedge product has an orientation.

We define the set of all these products to be *basic 2-forms*, and similarly we define the set of products of the form $dx^a \wedge dx^b \wedge dx^c$ to be *basic 3-forms*. A general k-form is then a weighted sum of basic k-forms, where the weights are the smooth functions f. Together these form a vector space with basic k-forms as the basis vectors, and 0-forms (smooth functions) as the field of scalars. The wedge product then extends to k-forms in the natural way. Over R^n at most n covectors can be linearly independent, thus a k-form with $k > n$ will always be zero, by the alternating property.

In addition to the wedge product, there is also the exterior derivative operator d. This operator maps k-forms to $(k+1)$-forms. For a k-form $\omega = f\,dx^a$ over R^n, we define the action of d by:

$$d\omega = \sum_{i=1}^{n} \frac{\partial f}{\partial x_i} dx^i \wedge dx^a.$$

with extension to general k-forms occurring linearly.

This more general approach allows for a more natural coordinate-free approach to integration on manifolds. It also allows for a natural generalisation of the fundamental theorem of calculus, called Stokes' theorem, which we may state as

$$\int_{\Omega} d\omega = \int_{\partial\Omega} \omega$$

where ω is a general k-form, and $\partial\Omega$ denotes the boundary of the region Ω. Thus, in the case that ω is a 0-form and Ω is a closed interval of the real line, this reduces to the fundamental theorem of calculus. In the case that ω is a 1-form and Ω is a two-dimensional region in the plane, the theorem reduces to Green's theorem. Similarly, using 2-forms, and 3-forms and Hodge duality, we can arrive at Stokes' theorem and the divergence theorem. In this way we can see that differential forms provide a powerful unifying view of integration.

Summations

The discrete equivalent of integration is summation. Summations and integrals can be put on the same foundations using the theory of Lebesgue integrals or time scale calculus.

Methods for Computing Integrals

Analytical: The most basic technique for computing definite integrals of one real variable is based on the fundamental theorem of calculus. Let $f(x)$ be the function of x to be integrated over a given interval $[a, b]$. Then, find an antiderivative of f; that is, a function F such that $F' = f$ on the interval. Provided the integrand and integral have no singularities on the path of integration, by the fundamental theorem of calculus, $\int_a^b f(x)dx = F(b) - F(a)$.

The integral is not actually the antiderivative, but the fundamental theorem provides a way to use antiderivatives to evaluate definite integrals.

The most difficult step is usually to find the antiderivative of f. It is rarely possible to glance at a function and write down its antiderivative. More often, it is necessary to use one of the many techniques that have been developed to evaluate integrals. Most of these techniques rewrite one integral as a different one which is hopefully more tractable. Techniques include:

- Integration by substitution
- Integration by parts
- Changing the order of integration
- Integration by trigonometric substitution
- Integration by partial fractions
- Integration by reduction formulae

- Integration using parametric derivatives
- Integration using Euler's formula
- Differentiation under the integral sign
- Contour integration

Alternate methods exist to compute more complex integrals. Many nonelementary integrals can be expanded in a Taylor series and integrated term by term. Occasionally, the resulting infinite series can be summed analytically. The method of convolution using Meijer G-functions can also be used, assuming that the integrand can be written as a product of Meijer G-functions. There are also many less common ways of calculating definite integrals; for instance, Parseval's identity can be used to transform an integral over a rectangular region into an infinite sum. Occasionally, an integral can be evaluated by a trick.

Computations of volumes of solids of revolution can usually be done with disk integration or shell integration. Specific results which have been worked out by various techniques are collected in the list of integrals.

Symbolic

Many problems in mathematics, physics, and engineering involve integration where an explicit formula for the integral is desired. Extensive tables of integrals have been compiled and published over the years for this purpose. With the spread of computers, many professionals, educators, and students have turned to computer algebra systems that are specifically designed to perform difficult or tedious tasks, including integration. Symbolic integration has been one of the motivations for the development of the first such systems, like Macsyma.

A major mathematical difficulty in symbolic integration is that in many cases, a closed formula for the antiderivative of a rather simple-looking function does not exist. For instance, it is known that the antiderivatives of the functions $\exp(x^2)$, x^x and $(\sin x)/x$ cannot be expressed in the closed form involving only rational and exponential functions, logarithm, trigonometric and inverse trigonometric functions, and the operations of multiplication and composition; in other words, none of the three given functions is integrable in elementary functions, which are the functions which may be built from rational functions, roots of a polynomial, logarithm, and exponential functions. The Risch algorithm provides a general criterion to determine whether the antiderivative of an elementary function is elementary, and, if it is, to compute it. Unfortunately, it turns out that functions with closed expressions of antiderivatives are the exception rather than the rule. Consequently, computerized algebra systems have no hope of being able to find an antiderivative for a randomly constructed elementary function. On the positive side, if the 'building blocks' for antiderivatives are fixed in advance, it may be still be possible to decide whether the antiderivative of a given function can be expressed using these blocks and operations of multiplication and composition, and to find the symbolic answer whenever it exists.

The Risch algorithm, implemented in Mathematica and other computer algebra systems, does just that for functions and antiderivatives built from rational functions, radicals, logarithm, and exponential functions. Some special integrands occur often enough to warrant special study. In particular, it may be useful to have, in the set of antiderivatives, the special functions of physics (like the Legendre functions, the hypergeometric function, the Gamma function, the Incomplete Gamma function and so on). Extending the Risch's algorithm to include such functions is possible but challenging and has been an active research subject.

More recently a new approach has emerged, using D-finite function, which are the solutions of linear differential equations with polynomial coefficients. Most of the elementary and special functions are D-finite and the integral of a D-finite function is also a D-finite function. This provide an algorithm to express the antiderivative of a D-finite function as the solution of a differential equation. This theory allows also to compute a definite integrals of a D-function as the sum of a series given by the first coefficients and an algorithm to compute any coefficient.

Some Important Definite Integrals

Mathematicians have used definite integrals as a tool to define identities. Among these identities is the definition of the Euler–Mascheroni constant:

$$\gamma = \int_1^{\infty}\left(\frac{1}{\lfloor x \rfloor} - \frac{1}{x}\right)dx,$$

the Gamma function:

$$\Gamma(z) = \int_0^{\infty} e^{-t}t^{z-1}dt,$$

and the Laplace transform which is widely used in engineering:

$$F(s) = \int_0^{\infty} e^{-st}f(t)dt.$$

Chapter 8

Regression Analysis

In statistics, regression analysis is a statistical technique for estimating the relationships among variables. It includes many techniques for modelling and analyzing several variables, when the focus is on the relationship between a dependent variable and one or more independent variables.

More specifically, regression analysis helps one understand how the typical value of the dependent variable changes when any one of the independent variables is varied, while the other independent variables are held fixed.

Most commonly, regression analysis estimates the conditional expectation of the dependent variable given the independent variables – that is, the average value of the dependent variable when the independent variables are fixed. Less commonly, the focus is on a quantile, or other location parameter of the conditional distribution of the dependent variable given the independent variables.

In all cases, the estimation target is a function of the independent variables called the regression function. In regression analysis, it is also of interest to characterize the variation of the dependent variable around the regression function, which can be described by a probability distribution.

Regression analysis is widely used for prediction and forecasting, where its use has substantial overlap with the field of machine learning. Regression analysis is also used to understand which among the independent variables are related to the dependent variable, and to explore the forms of these relationships.

In restricted circumstances, regression analysis can be used to infer causal relationships between the independent and dependent variables. However this can lead to illusions or false relationships, so caution is advisable; for example, correlation does not imply causation.

A large body of techniques for carrying out regression analysis has been developed. Familiar methods such as linear regression and ordinary least squares regression are parametric, in that the regression function is defined in terms of a finite number of unknown parameters that are estimated from the data. Nonparametric regression refers to techniques that allow the regression function to lie in a specified set of functions, which may be infinite-dimensional.

The performance of regression analysis methods in practice depends on the form of the data generating process, and how it relates to the regression approach being used. Since the true form of the data-generating process is generally not known, regression analysis often depends to some extent on making assumptions about this process.

These assumptions are sometimes testable if many data are available. Regression models for prediction are often useful even when the assumptions are moderately violated, although they may not perform optimally. However, in many applications, especially with small effects or questions of causality based on observational data, regression methods can give misleading results.

History

The earliest form of regression was the method of least squares, which was published by Legendre in 1805, and by Gauss in 1809. Legendre and Gauss both applied the method to the problem of determining, from astronomical observations, the orbits of bodies about the Sun (mostly comets, but also later the then newly discovered minor planets). Gauss published a further development of the theory of least squares in 1821, including a version of the Gauss–Markov theorem.

The term "regression" was coined by Francis Galton in the nineteenth century to describe a biological phenomenon. The phenomenon was that the heights of descendants of tall ancestors tend to regress down towards a normal average (a phenomenon also known as regression toward the mean).

For Galton, regression had only this biological meaning, but his work was later extended by Udny Yule and Karl Pearson to a more general statistical context.

In the work of Yule and Pearson, the joint distribution of the response and explanatory variables is assumed to be Gaussian. This assumption was weakened by R.A. Fisher in his works of 1922 and 1925. Fisher assumed that the conditional distribution of the response variable is Gaussian, but the joint distribution need not be. In this respect, Fisher's assumption is closer to Gauss's formulation of 1821.

In the 1950s and 1960s, economists used electromechanical desk calculators to calculate regressions. Before 1970, it sometimes took up to 24 hours to receive the result from one regression.

Regression methods continue to be an area of active research. In recent decades, new methods have been developed for robust regression, regression involving correlated responses such as time series and growth curves, regression in which the predictor or response variables are curves, images, graphs, or other complex data objects, regression methods accommodating various types of missing data, nonparametric regression, Bayesian methods for regression, regression in which the predictor variables are measured with error, regression with more predictor variables than observations, and causal inference with regression.

Regression Models

Regression models involve the following variables:

- The unknown parameters, denoted as β, which may represent a scalar or a vector.
- The independent variables, X.
- The dependent variable, Y.

In various fields of application, different terminologies are used in place of dependent and independent variables.

A regression model relates Y to a function of X and β.

$$Y \approx f(X,\beta)$$

The approximation is usually formalized as $E(Y \mid X) = f(X, \beta)$. To carry out regression analysis, the form of the function f must be specified. Sometimes the form of this function is based on knowledge about the relationship between Y and X that does not rely on the data. If no such knowledge is available, a flexible or convenient form for f is chosen.

Assume now that the vector of unknown parameters β is of length k. In order to perform a regression analysis the user must provide information about the dependent variable Y:

- If N data points of the form (Y,X) are observed, where $N < k$, most classical approaches to regression analysis cannot be performed: since the system of equations defining the regression model is underdetermined, there are not enough data to recover β.
- If exactly $N = k$ data points are observed, and the function f is linear, the equations $Y = f(X, \beta)$ can be solved exactly rather than approximately. This reduces to solving a set of N equations with N unknowns (the elements of β), which has a unique solution as long as the X are linearly independent. If f is nonlinear, a solution may not exist, or many solutions may exist.
- The most common situation is where $N > k$ data points are observed. In this case, there is enough information in the data to estimate a unique value for β that best fits the data in some sense, and the regression model when applied to the data can be viewed as an overdetermined system in β.

In the last case, the regression analysis provides the tools for:

1. Finding a solution for unknown parameters β that will, for example, minimize the distance between the measured and predicted values of the dependent variable Y (also known as method of least squares).
2. Under certain statistical assumptions, the regression analysis uses the surplus of information to provide statistical information about the unknown parameters β and predicted values of the dependent variable Y.

Necessary Number of Independent Measurements

Consider a regression model which has three unknown parameters, β_0, β_1, and β_2. Suppose an experimenter performs 10 measurements all at exactly the same value of independent variable vector X (which contains the independent variables X_1, X_2, and X_3). In this case, regression analysis fails to give a unique set of estimated values for the three unknown parameters; the experimenter did not provide enough information. The best one can do is to estimate the average value and the standard deviation of the dependent variable Y. Similarly, measuring at two different values of X would give enough data for a regression with two unknowns, but not for three or more unknowns.

If the experimenter had performed measurements at three different values of the independent variable vector X, then regression analysis would provide a unique set of estimates for the three unknown parameters in β.

In the case of general linear regression, the above statement is equivalent to the requirement that the matrix X^TX is invertible.

Statistical Assumptions

When the number of measurements, N, is larger than the number of unknown parameters, k, and the measurement errors ε_i are normally distributed then *the excess of information* contained in $(N - k)$ measurements is used to make statistical predictions about the unknown parameters. This excess of information is referred to as the degrees of freedom of the regression.

Underlying Assumptions

Classical assumptions for regression analysis include:

- The sample is representative of the population for the inference prediction.
- The error is a random variable with a mean of zero conditional on the explanatory variables.
- The independent variables are measured with no error. (Note: If this is not so, modelling may be done instead using errors-in-variables model techniques).
- The predictors are linearly independent, i.e. it is not possible to express any predictor as a linear combination of the others.
- The errors are uncorrelated, that is, the variance–covariance matrix of the errors is diagonal and each non-zero element is the variance of the error.
- The variance of the error is constant across observations (homoscedasticity). If not, weighted least squares or other methods might instead be used.

These are sufficient conditions for the least-squares estimator to possess desirable properties; in particular, these assumptions imply that the parameter estimates will be unbiased, consistent, and efficient in the class of linear unbiased estimators. It is important to note that actual data rarely satisfies the assumptions.

That is, the method is used even though the assumptions are not true. Variation from the assumptions can sometimes be used as a measure of how far the model is from being useful. Many of these assumptions may be relaxed in more advanced treatments. Reports of statistical analyses usually include analyses of tests on the sample data and methodology for the fit and usefulness of the model.

Assumptions include the geometrical support of the variables. Independent and dependent variables often refer to values measured at point locations. There may be spatial trends and spatial autocorrelation in the variables that violates statistical assumptions of regression. Geographic weighted regression is one technique to deal with such data. Also, variables may include values aggregated by areas. With aggregated data the modifiable areal unit problem can cause extreme variation in regression parameters. When analyzing data aggregated by political boundaries, postal codes or census areas results may be very different with a different choice of units.

Linear Regression

In linear regression, the model specification is that the dependent variable, y_i is a linear combination of the *parameters* (but need not be linear in the *independent variables*). For example, in simple linear regression for modelling *n* data points there is one independent variable: x_i, and two parameters, β_0 and β_1:

straight line: $y_i = \beta_0 + \beta_1 x_i + \varepsilon_i, \quad i = 1, \ldots, n.$

In multiple linear regression, there are several independent variables or functions of independent variables.

Adding a term in x_i^2 to the preceding regression gives:

parabola: $y_i = \beta_0 + \beta_1 x_i + \beta_2 x_i^2 + \varepsilon_i, i = 1, \ldots, n.$

This is still linear regression; although the expression on the right hand side is quadratic in the independent variable x_i, it is linear in the parameters β_0, β_1 and β_2. In both cases, ε_i is an error term and the subscript *i* indexes a particular observation. Given a random sample from the population, we estimate the population parameters and obtain the sample linear regression model:

$$\widehat{y_i} = \hat{\beta}_0 + \hat{\beta}_1 x_i.$$

The residual, $e_i = y_i - \hat{y}_i$, is the difference between the value of the dependent variable predicted by the model, y_i, and the true value of the dependent variable, y_i. One method of estimation is ordinary least squares. This method obtains parameter estimates that minimize the sum of squared residuals, SSE, also sometimes denoted RSS:

$$SSE = \sum_{i=1}^{n} e_i^2.$$

Minimization of this function results in a set of normal equations, a set of simultaneous linear equations in the parameters, which are solved to yield the parameter estimators, $\hat{\beta}_0, \hat{\beta}_1$.

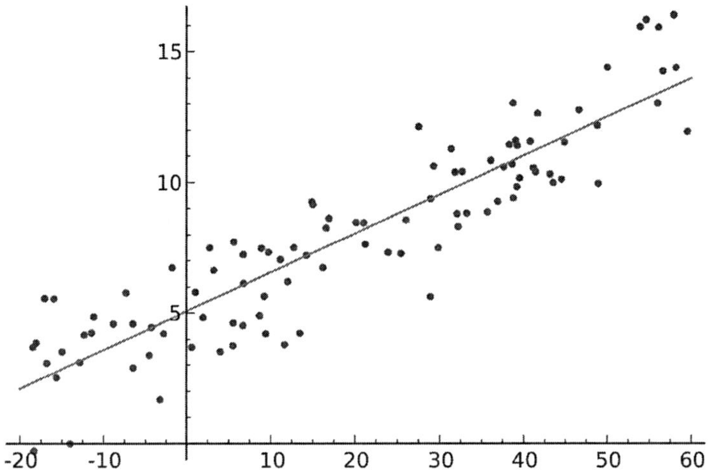

Figure: *Illustration of linear regression on a data set.*

In the case of simple regression, the formulas for the least squares estimates are

$$\widehat{\beta_1} = \frac{\sum(x_i - \overline{x})(y_i - \overline{y})}{\sum(x_i - \overline{x})^2} \text{ and } \widehat{\beta_0} = \overline{y} - \widehat{\beta_1}\overline{x}$$

where $\overline{x}$ is the mean (average) of the x values and $\overline{y}$ is the mean of the y values. Under the assumption that the population error term has a constant variance, the estimate of that variance is given by:

$$\hat{\sigma}_\varepsilon^2 = \frac{SSE}{n-2}.$$

This is called the mean square error (MSE) of the regression. The denominator is the sample size reduced by the number of model parameters estimated from the same data, (n-p) for p regressors or (n-p-1) if an intercept is used. In this case, p=1 so the denominator is n-2. The standard errors of the parameter estimates are given by

$$\hat{\sigma}_{\beta_0} = \hat{\sigma}_\varepsilon \sqrt{\frac{1}{n} + \frac{\overline{x}^2}{\sum(x_i - \overline{x})^2}}$$

$$\hat{\sigma}_{\beta_1} = \hat{\sigma}_\varepsilon \sqrt{\frac{1}{\sum(x_i - \overline{x})^2}}.$$

Under the further assumption that the population error term is normally distributed, the researcher can use these estimated standard errors to create confidence intervals and conduct hypothesis tests about the population parameters.

General Linear Model

In the more general multiple regression model, there are p independent variables:

$$y_i = \beta_1 x_{i1} + \beta_2 x_{i2} + \cdots + \beta_p x_{ip} + \varepsilon_i,$$

where x_{ij} is the ith observation on the jth independent variable, and where the first independent variable takes the value 1 for all i (so β_1 is the regression intercept).

The least squares parameter estimates are obtained from p normal equations. The residual can be written as

$$\varepsilon_i = y_i - \hat{\beta}_1 x_{i1} - \cdots - \hat{\beta}_p x_{ip}.$$

The normal equations are

$$\sum_{i=1}^{n}\sum_{k=1}^{p} X_{ij} X_{ik} \hat{\beta}_k = \sum_{i=1}^{n} X_{ij} y_i,\ j = 1, \ldots, p.$$

In matrix notation, the normal equations are written as

$$(\mathbf{X}^\top \mathbf{X})\hat{\beta} = \mathbf{X}^\top \mathbf{Y},$$

where the ij element of X is x_{ij}, the i element of the column vector Y is y_i, and the j element of $\hat{\beta}$ is $\hat{\beta}_j$. Thus X is $n\times p$, Y is $n\times 1$, and is $p\times 1$. The solution is

$$\hat{\beta} = (\mathbf{X}^\top \mathbf{X})^{-1} \mathbf{X}^\top \mathbf{Y}.$$

Diagnostics

Once a regression model has been constructed, it may be important to confirm the goodness of fit of the model and the statistical significance of the estimated parameters. Commonly used checks of goodness of fit include the R-squared, analyses of the pattern of residuals and hypothesis testing. Statistical significance can be checked by an F-test of the overall fit, followed by t-tests of individual parameters.

Interpretations of these diagnostic tests rest heavily on the model assumptions. Although examination of the residuals can be used to invalidate a model, the results of a t-test or F-test are sometimes more difficult to interpret if the model's assumptions are violated. For example, if the error term does not have a normal distribution, in small samples the estimated parameters will not follow normal distributions and complicate inference. With relatively large samples, however, a central limit theorem can be invoked such that hypothesis testing may proceed using asymptotic approximations.

"Limited Dependent" Variables

The phrase "limited dependent" is used in econometric statistics for categorical and constrained variables.

The response variable may be non-continuous ("limite≤ to lie on some subset of the real line). For binary (zero or one) variables, if analysis proceeds with least-squares linear regression, the model is called the linear probability model. Nonlinear models for binary dependent variables include the probit and logit model. The multivariate probit model is a standard method of estimating a joint relationship between several binary dependent variables and some independent variables. For categorical variables with more than two values there is the multinomial logit. For ordinal variables with more than two values, there are the ordered logit and ordered probit models. Censored regression models may be used when the dependent variable is only sometimes observed, and Heckman correction type models may be used when the sample is not randomly selected from the population of interest. An alternative to such procedures is linear regression based on polychoric correlation (or polyserial correlations) between the categorical variables. Such procedures differ in the assumptions made about the distribution of the variables in the population. If the variable is positive with low values and represents the repetition of the occurrence of an event, then count models like the Poisson regression or the negative binomial model may be used instead.

Interpolation and Extrapolation

Regression models predict a value of the Y variable given known values of the X variables. Prediction *within* the range of values in the dataset used for model-fitting is known informally as interpolation. Prediction *outside* this range of the data is known as extrapolation. Performing extrapolation relies strongly on the regression assumptions. The further the extrapolation goes outside the data, the more room there is for the model to fail due to differences between the assumptions and the sample data or the true values.

It is generally advised that when performing extrapolation, one should accompany the estimated value of the dependent variable with a prediction interval that represents the uncertainty. Such intervals tend to expand rapidly as the values of the independent variable(s) moved outside the range covered by the observed data. For such reasons and others, some tend to say that it might be unwise to undertake extrapolation.

However, this does not cover the full set of modelling errors that may be being made: in particular, the assumption of a particular form for the relation between Y and X. A properly conducted regression analysis will include an assessment of how well the assumed form is matched by the observed data, but it can only do so within the range of values of the independent variables actually available. This means that any extrapolation is particularly reliant on the assumptions being made about the structural form of the regression relationship. Best-practice advice here is that a linear-in-variables and linear-in-parameters relationship should not be chosen simply for computational convenience, but that all available knowledge should be deployed in constructing a regression model. If this knowledge includes the fact that the dependent variable cannot go outside a certain range of values, this can be made use of in selecting the model – even if the observed dataset has no values particularly near such bounds. The implications of this step of choosing an appropriate functional form for the regression can be great when extrapolation is considered. At a minimum, it can ensure that any extrapolation arising from a fitted model is "realistic" (or in accord with what is known).

Nonlinear Regression

When the model function is not linear in the parameters, the sum of squares must be minimized by an iterative procedure. This introduces many complications which are summarized in Differences between linear and non-linear least squares

Power and Sample Size Calculations

There are no generally agreed methods for relating the number of observations versus the number of independent variables in the model. One rule of thumb suggested by Good and Hardin is $N = m^n$, where N is the sample size, n is the number of independent variables and m is the number of observations needed to reach the desired precision if the model had only one independent variable. For example, a researcher is building a linear regression model using a dataset that contains 1000 patients (N). If he decides that five observations are needed to precisely define a straight line (m), then the maximum number of independent variables his model can support is 4, because

$$\frac{\log 1000}{\log 5} = 4.29.$$

Other Methods

Although the parameters of a regression model are usually estimated using the method of least squares, other methods which have been used include:

- Bayesian methods, e.g. Bayesian linear regression
- Percentage regression, for situations where reducing *percentage* errors is deemed more appropriate.
- Least absolute deviations, which is more robust in the presence of outliers, leading to quantile regression
- Nonparametric regression, requires a large number of observations and is computationally intensive
- Distance metric learning, which is learned by the search of a meaningful distance metric in a given input space.

Software

All major statistical software packages perform least squares regression analysis and inference. Simple linear regression and multiple regression using least squares can be done in some spreadsheet applications and on some calculators. While many statistical software packages can perform various types of nonparametric and robust regression, these methods are less standardized; different software packages implement different methods, and a method with a given name may be implemented differently in different packages. Specialized regression software has been developed for use in fields such as survey analysis and neuroimaging.

Simple Linear Regression

In statistics, simple linear regression is the least squares estimator of a linear regression model with a single explanatory variable. In other words, simple linear regression fits a straight line through the set of n points in such a way that makes the sum of squared *residuals* of the model (that is, vertical distances between the points of the data set and the fitted line) as small as possible.

The adjective *simple* refers to the fact that this regression is one of the simplest in statistics. The slope of the fitted line is equal to the correlation between y and x corrected by the ratio of standard deviations of these variables. The intercept of the fitted line is such that it passes through the centre of mass (x, y) of the data points.

Other regression methods besides the simple ordinary least squares (OLS) also exist. In particular, when one wants to do regression by eye, people usually tend to draw a slightly steeper line, closer to the one produced by the total least squares method.

This occurs because it is more natural for one's mind to consider the orthogonal distances from the observations to the regression line, rather than the vertical ones as OLS method does.

Fitting the Regression Line

Suppose there are n data points $\{y_i, x_i\}$, where $i = 1, 2, \ldots, n$. The goal is to find the equation of the straight line

$$y = \alpha + \beta x,$$

which would provide a "best" fit for the data points. Here the "best" will be understood as in the least-squares approach: such a line that minimizes the sum of squared residuals of the linear regression model. In other words, numbers α and β solve the following minimization problem:

$$\text{Find } \min_{\alpha,\beta} Q(\alpha,\beta), \text{ where } Q(\alpha,\beta) = \sum_{i=1}^{n} \hat{\varepsilon}_i^2 = \sum_{i=1}^{n} (y_i - \alpha - \beta x_i)^2$$

By using either calculus, the geometry of inner product spaces or simply expanding to get a quadratic in α and β, it can be shown that the values of α and $\hat{a}$ that minimize the objective function Q are

$$\hat{\beta} \quad = \frac{\sum_{i=1}^{n}(x_i - \bar{x})(y_i - \bar{y})}{\sum_{i=1}^{n}(x_i - \bar{x})^2} = \frac{\sum_{i=1}^{n} x_i y_i - \frac{1}{n}\sum_{i=1}^{n} x_i \sum_{j=1}^{n} y_j}{\sum_{i-1}^{n}(x_i^2) - \frac{1}{n}(\sum_{i=1}^{n} x_i)^2}$$

$$[6pt] \quad = \frac{\overline{xy} - \bar{x}\bar{y}}{\overline{x^2} - \bar{x}^2} = \frac{\mathrm{Cov}[x,y]}{\mathrm{Var}[x]} = r_{xy}\frac{s_y}{s_x},$$

$$[6pt]\hat{\alpha} \quad = \bar{y} - \hat{\beta}\bar{x},$$

where r_{xy} is the sample correlation coefficient between x and y, s_x is the standard deviation of x, and s_y is correspondingly the standard deviation of y.

Horizontal bar over a variable means the sample average of that variable. For example: $\overline{xy} = \frac{1}{n}\sum_{i=1}^{n} x_i y_i$.

Substituting the above expressions for $\hat{\alpha}$ and $\hat{\beta}$ into

$$y = \hat{\alpha} + \hat{\beta}x,$$

yields

$$\frac{y - \overline{y}}{s_y} = r_{xy}\frac{x - \overline{x}}{s_x}$$

This shows the role r_{xy} plays in the regression line of standardized data points.

Linear Regression without the Intercept Term

Sometimes, people consider a simple linear regression model without the intercept term: $y = \beta x$. In such a case, the OLS estimator for β simplifies to $\hat{\beta} = (\overline{xy}) / (\overline{x^2})$. ·

Numerical Properties

1. The line goes through the "centre of mass" point (x, y).
2. The sum of the residuals is equal to zero, if the model includes a constant: $\sum_{i=1}^{n} \hat{\varepsilon}_i = 0$.
3. The linear combination of the residuals, in which the coefficients are the x-values, is equal to zero: $\sum_{i=1}^{n} x_i\hat{\varepsilon}_i = 0$.

Model-cased Properties

Description of the statistical properties of estimators from the simple linear regession estimates requires the use of a statistical model. The following is based on assuming the validity of a model under which the estimates are optimal. It is also possible to evaluate the properties under other assumptions, such as inhomogeneity, but this is discussed elsewhere.

Unbiasedness

The estimators $\hat{\alpha}$ and $\hat{\beta}$ are unbiased. This requires that we interpret the estimators as random variables and so we have to assume that, for each value of x, the corresponding value of y is generated as a mean response $\alpha + \beta x$ plus an additional random variable ε called the *error term*. This error term has to be equal to zero on average, for each value of x. Under such interpretation, the least-squares estimators and will themselves be random variables, and they will unbiasedly estimate the "true values" α and β.

Confidence Intervals

The formulas given in the previous section allow one to calculate the *point estimates* of α and β — that is, the coefficients of the regression line for the given set of data. However, those formulas do not tell us how precise the estimates are. That is, how much the estimators $\hat{\alpha}$ and $\hat{\beta}$ can deviate from the "true" values of α and β. The latter question is answered by the *confidence intervals* for the regression coefficients.

In order to construct the confidence intervals usually one of the two possible assumptions is made: either that the errors in the regression are normally distributed (the so-called *classic regression* assumption), or that the number of observations n is sufficiently large so that the actual distribution of the estimators can be approximated using the central limit theorem.

Normality Assumption

Under the first assumption above, that of the normality of the error terms, the estimator of the slope coefficient will itself be normally distributed with mean β and variance $\sigma^2 / \sum(x_i - \bar{x})^2$, where σ^2 is the variance of the error terms. At the same time the sum of squared residuals Q is distributed proportionally to χ^2 with $(n-2)$ degrees of freedom, and independently from $\hat{\beta}$. This allows us to construct a t-statistic

$$t = \frac{\hat{\beta} - \beta}{s_{\hat{\beta}}} \sim t_{n-2}, \quad \text{where} \quad s_{\hat{\beta}} = \sqrt{\frac{\frac{1}{n-2}\sum_{i=1}^{n}\hat{\varepsilon}_i^2}{\sum_{i=1}^{n}(x_i - \bar{x})^2}}$$

which has a Student's t-distribution with $(n-2)$ degrees of freedom. Here s_β is the *standard error* of the estimator $\hat{\beta}$. Using this t-statistic we can construct a confidence interval for β:

$$:\beta \in \left[\hat{\beta} - s_{\hat{\beta}} t^*_{n-2}, \hat{\beta} + s_{\hat{\beta}} t^*_{n-2}\right] \text{ at confidence level } (1-\gamma),$$

where t^*_{n-2} is the $(1-\gamma/2)$-th quantile of the t_{n-2} distribution. For example, if $\gamma = 0.05$ then the confidence level is 95%. Similarly, the confidence interval for the intercept coefficient α is given by

$$\alpha \in \left[\hat{\alpha} - s_{\hat{\alpha}} t^*_{n-2}, \hat{\alpha} + s_{\hat{\alpha}} t^*_{n-2}\right] \text{ at confidence level } (1-\gamma),$$

where

$$s_{\hat{\alpha}} = s_{\hat{\beta}}\sqrt{\tfrac{1}{n}\sum\nolimits_{i=1}^{n} x_i^2} = \sqrt{\frac{1}{n(n-2)}\left(\sum\nolimits_{j=1}^{n}\hat{\varepsilon}_j^2\right)\frac{\sum_{i=1}^{n} x_i^2}{\sum_{i=1}^{n}(x_i-\overline{x})^2}}$$

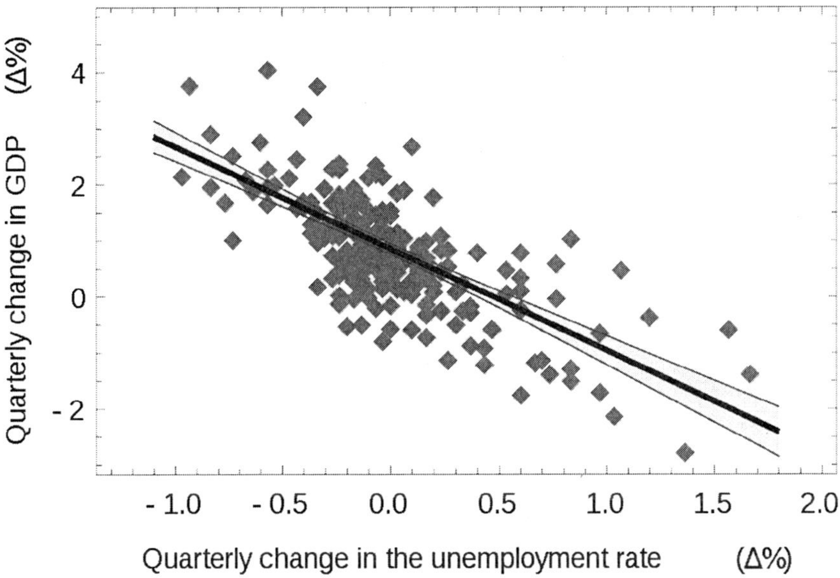

Figure: *The US "changes in unemployment – GDP growth" regression with the 95% confidence bands.*

The confidence intervals for α and β give us the general idea where these regression coefficients are most likely to be. For example in the "Okun's law" regression shown at the beginning of the article the point estimates are $\hat{\alpha} = 0.859$ and $\hat{\beta} = -1.817$. The 95% confidence intervals for these estimates are

$$\alpha \in [0.76, 0.96], \quad \beta \in [-2.06, -1.58] \text{ with 95\% confidence.}$$

In order to represent this information graphically, in the form of the confidence bands around the regression line, one has to proceed carefully and account for the joint distribution of the estimators. It can be shown that at confidence level $(1-\gamma)$ the confidence band has hyperbolic form given by the equation

$$\hat{y}\,|_{x=\xi} \in \left[\hat{\alpha} + \hat{\beta}\xi \pm t_{n-2}^{*}\sqrt{\frac{1}{n-2}\sum\hat{\varepsilon}_i^2\cdot\left(\frac{1}{n} + \frac{(\xi-\overline{x})^2}{\sum(x_i-\overline{x})^2}\right)}\right].$$

Asymptotic Assumption

The alternative second assumption states that when the number of points in the dataset is "large enough", the law of large numbers and the central limit theorem become applicable, and then the distribution of the estimators is approximately normal. Under this assumption all formulas derived in the previous section remain valid, with the only exception that the quantile t^*_{n-2} of student-t distribution is replaced with the quantile q^* of the standard normal distribution. Occasionally the fraction $^1D_{(n-2)}$ is replaced with ^1D_n. When n is large such change does not alter the results considerably.

Numerical Example

This example concerns the data set from the Ordinary least squares article. This data set gives average weights for humans as a function of their height in the population of American women of age 30–39. Although the OLS article argues that it would be more appropriate to run a quadratic regression for this data, the simple linear regression model is applied here instead.

x_i	1.47	1.50	1.52	1.55	1.57	1.60	1.63	1.65	1.68	1.70	1.73	1.75	1.78	1.80	1.83	**Height (m)**
y_i	52.21	53.12	54.48	55.84	57.20	58.57	59.93	61.29	63.11	64.47	66.28	68.10	69.92	72.19	74.46	**Mass (kg)**

There are $n = 15$ points in this data set. Hand calculations would be started by finding the following five sums:

$$S_x = \sum x_i = 24.76, \quad S_y = \sum y_i = 931.17$$

$$S_{xx} = \sum x_i^2 = 41.0532, \quad S_{xy} = \sum x_i y_i = 1548.2453, \quad S_{yy} = \sum y_i^2 = 58498.5439$$

These quantities would be used to calculate the estimates of the regression coefficients, and their standard errors.

$$\hat{\beta} = \frac{nS_{xy} - S_x S_y}{nS_{xx} - S_x^2} = 61.272$$

$$\hat{\alpha} = \tfrac{1}{n} S_y - \hat{\beta} \tfrac{1}{n} S_x = -39.062$$

$$s_\varepsilon^2 = \tfrac{1}{n(n-2)}\left(nS_{yy} - S_y^2 - \hat{\beta}^2 (nS_{xx} - S_x^2)\right) = 0.5762$$

$$s_\beta^2 = \frac{n s_\varepsilon^2}{nS_{xx} - S_x^2} = 3.1539$$

$$s_\alpha^2 = s_\beta^2 \tfrac{1}{n} S_{xx} = 8.63185$$

The 0.975 quantile of Student's t-distribution with 13 degrees of freedom is $t^*_{13} = 2.1604$, and thus confidence intervals for α and β are

$$\alpha \in [\hat{\alpha} \mp t^*_{13} s_\alpha] = [-45.4, -32.7]$$

$$\beta \in [\hat{\beta} \mp t^*_{13} s_\beta] = [57.4, 65.1]$$

The product-moment correlation coefficient might also be calculated:

$$\hat{r} = \frac{nS_{xy} - S_x S_y}{\sqrt{(nS_{xx} - S_x^2)(nS_{yy} - S_y^2)}} = 0.9945$$

This example also demonstrates that sophisticated calculations will not overcome the use of badly prepared data. The heights were originally given in inches, and have been converted to the nearest centimetre. Since the conversion factor is one inch to 2.54 cm, this is *not* a correct conversion. The original inches can be recovered by Round(x/0.0254) and then re-converted to metric: if this is done, the results become

$$\hat{\beta} = 61.6746$$

$$\hat{\alpha} = -39.7468$$

Thus a seemingly small variation in the data has a real effect.

Bayesian Multivariate Linear Regression

In statistics, Bayesian multivariate linear regression is a Bayesian approach to multivariate linear regression, i.e. linear regression where the predicted outcome is a vector of correlated random variables rather than a single scalar random variable.

Details

Consider a regression problem where the dependent variable to be predicted is not a single real-valued scalar but an m-length vector of correlated real numbers. As in the standard regression setup, there are n observations, where each observation i consists of k-1 explanatory variables, grouped into a vector $\mathbf{x}_i$ -of length k (where a dummy variable with a value of 1 has been added to allow for an intercept coefficient). This can be viewed as a set of m related regression problems for each observation i:

$$y_{i,1} = \mathbf{x}_i^{\mathrm{T}} \beta_1 + \epsilon_{i,1}$$

$$\cdots$$

$$y_{i,m} = \mathbf{x}_i^{\mathrm{T}} \beta_m + \epsilon_{i,m}$$

where the set of errors $\{\epsilon_{i,1},\ldots,\epsilon_{i,m}\}$ are all correlated. Equivalently, it can be viewed as a single regression problem where the outcome is a row vector y_i^T and the regression coefficient vectors are stacked next to each other, as follows:

$$y_i^T = x_i^T B + \epsilon_i^T .$$

The coefficient matrix B is a $k \times m$ matrix where the coefficient vectors $\beta_1,\ldots,\beta_m$ for each regression problem are stacked horizontally:

$$\mathbf{B} = \left[\begin{pmatrix} \\ \beta_1 \\ \\ \end{pmatrix} \cdots \begin{pmatrix} \\ \beta_m \\ \\ \end{pmatrix} \right] = \left[\begin{pmatrix} \beta_{1,1} \\ \vdots \\ \beta_{1,k} \end{pmatrix} \cdots \begin{pmatrix} \beta_{m,1} \\ \vdots \\ \beta_{m,k} \end{pmatrix} \right].$$

The noise vector ϵ_i for each observation i is jointly normal, so that the outcomes for a given observation are correlated:

$$\epsilon_i \sim N(0, \Sigma_\epsilon^2).$$

We can write the entire regression problem in matrix form as:

$$Y = XB + E,$$

where Y and E are $n \times m$ matrices. The design matrix X is an $n \times k$ matrix with the observations stacked vertically, as in the standard linear regression setup:

$$:\mathbf{X} = \begin{bmatrix} \mathbf{x}_1^T \\ \mathbf{x}_2^T \\ \vdots \\ \mathbf{x}_n^T \end{bmatrix} = \begin{bmatrix} x_{1,1} & \cdots & x_{1,k} \\ x_{2,1} & \cdots & x_{2,k} \\ \vdots & \ddots & \vdots \\ x_{n,1} & \cdots & x_{n,k} \end{bmatrix}.$$

The classical, frequentists linear least squares solution is to simply estimate the matrix of regression coefficients $\hat{B}$ using the Moore-Penrose pseudoinverse:

$$\hat{B} = (X^T X)^{-1} X^T Y \cdot$$

To obtain the Bayesian solution, we need to specify the conditional likelihood and then find the appropriate conjugate prior. As with the univariate case of linear Bayesian regression, we will find that we can specify a natural conditional conjugate prior (which is scale dependent).

Let us write our conditional likelihood as

$$\rho(\mathbf{E} \mid \Sigma_\epsilon) \propto (\Sigma_\epsilon^2)^{-n/2} \exp(-\frac{1}{2} \mathrm{tr}(\mathbf{E}^T \Sigma_\epsilon^{-1} \mathbf{E})),$$

writing the error E in terms of Y, X and B yields

$$\rho(Y \mid X, B, \Sigma_\epsilon) \propto (\Sigma_\epsilon^2)^{-n/2} \exp(-\frac{1}{2} tr((Y - XB)^T \Sigma_\epsilon^{-1} (Y - XB))),$$

We seek a natural conjugate prior—a joint density $\rho(\mathbf{B}, \sigma_\epsilon)$ which is of the same functional form as the likelihood. Since the likelihood is quadratic in , we re-write the likelihood so it is normal in (the deviation from classical sample estimate). Using the same technique as with Bayesian linear regression, we decompose the exponential term using a matrix-form of the sum-of-squares technique. Here, however, we will also need to use the Matrix Differential Calculus (Kronecker product and vectorization transformations).

First, let us apply sum-of-squares to obtain new expression for the likelihood:

$$\rho(Y \mid X, B, \Sigma_\epsilon) \propto \Sigma_\epsilon^{-(n-k)/2} \exp(-tr(-\frac{1}{2} S^T \Sigma_\epsilon^{-1} S))(\Sigma_\epsilon^2)^{-k/2} \exp(-\frac{1}{2} tr((B - \hat{B})^T X^T \Sigma_\epsilon^{-1} X (B - \hat{B}))),$$

$$S = Y - \hat{B}X$$

We would like to develop a conditional form for the priors:

$$\rho(B, \Sigma_\epsilon) = \rho(\Sigma_\epsilon) \rho(B \mid \Sigma_\epsilon),$$

where $\rho(\Sigma_\epsilon)$ is an inverse-Wishart distribution and $\rho(B \mid \Sigma_\epsilon)$ is some form of normal distribution in the matrix B. This is accomplished using the vectorization transformation, which converts the likelihood from a function of the matrices $B, \hat{B}$ to a function of the vectors $\beta = vec(B), \hat{\beta} = vec(\hat{B})$.

Write

$$tr((B - \hat{B})^T X^T \Sigma_\epsilon^{-1} X (B - \hat{B})) = vec(B - \hat{B})^T vec(X^T \Sigma_\epsilon^{-1} X (B - \hat{B}))$$

Let

$$vec(X^T \Sigma_\epsilon^{-1} X (B - \hat{B})) = (\Sigma_\epsilon^{-1} \otimes X^T X) vec(B - \hat{B}),$$

where $\mathbf{A} \otimes \mathbf{B}$ denotes the Kronecker product of matrices A and B, a generalization of the outer product which multiplies an $m \times n$ matrix by a $p \times q$ matrix to generate an $mp \times nq$ matrix, consisting of every combination of products of elements from the two matrices.

Then

$$vec(B - \hat{B})^T (\Sigma_\epsilon^{-1} \otimes X^T X) vec(B - \hat{B}) = (\beta - \hat{\beta})(\Sigma_\epsilon^{-1} \otimes X^T X)(\beta - \hat{\beta})$$

$$= (\beta - \hat{\beta})(\Sigma_\epsilon^{-1} \otimes X^T X)(\beta - \hat{\beta})$$

which will lead to a likelihood which is normal in $(\beta - \hat{\beta})$..

With the likelihood in a more tractable form, we can now find a natural (conditional) conjugate prior.

Ordinary Least Squares

In statistics, ordinary least squares (OLS) or linear least squares is a method for estimating the unknown parameters in a linear regression model. This method minimizes the sum of squared vertical distances between the observed responses in the dataset and the responses predicted by the linear approximation. The resulting estimator can be expressed by a simple formula, especially in the case of a single regressor on the right-hand side.

The OLS estimator is consistent when the regressors are exogenous and there is no perfect multicollinearity, and optimal in the class of linear unbiased estimators when the errors are homoscedastic and serially uncorrelated. Under these conditions, the method of OLS provides minimum-variance mean-unbiased estimation when the errors have finite variances. Under the additional assumption that the errors be normally distributed, OLS is the maximum likelihood estimator. OLS is used in economics (econometrics), Political Science and electrical engineering (control theory and signal processing), among many areas of application.

Linear Model

Suppose the data consists of n observations $\{\text{yi,xi}\}ni = 1$ Each observation includes a scalar response y_i and a vector of p predictors (or regressors) x_i. In a linear regression model the response variable is a linear function of the regressors:

$$y_i = x_i'\beta + \varepsilon_i,$$

where β is a $p\times1$ vector of unknown parameters; ε_i's are unobserved scalar random variables (errors) which account for the discrepancy between the actually observed responses y_i and the "predicted outcomes" $x2\ _i\beta$; and 2 denotes matrix transpose, so that $x2\ \beta$ is the dot product between the vectors x and β. This model can also be written in matrix notation as

$$y = X\beta + \varepsilon,$$

where y and ε are $n\times1$ vectors, and X is an $n\times p$ matrix of regressors, which is also sometimes called the design matrix. As a rule, the constant term is always included in the set of regressors X, say, by taking $x_{i1} = 1$ for all $i = 1, \ldots, n$. The coefficient β_1 corresponding to this regressor is called the *intercept*.

There may be some relationship between the regressors. For instance, the third regressor may be the square of the second regressor. In this case (assuming that the first regressor is constant) we have a quadratic model in the second regressor. But this is still considered a linear model because it is linear in the βs.

Assumptions

There are several different frameworks in which the linear regression model can be cast in order to make the OLS technique applicable. Each of these settings produces the same formulas and same results, the only difference is the interpretation and the assumptions which have to be imposed in order for the method to give meaningful results. The choice of the applicable framework depends mostly on the nature of data at hand, and on the inference task which has to be performed.

One of the lines of difference in interpretation is whether to treat the regressors as random variables, or as predefined constants. In the first case (random design) the regressors x_i are random and sampled together with the y_i's from some population, as in an observational study. This approach allows for more natural study of the asymptotic properties of the estimators. In the other interpretation (fixed design), the regressors X are treated as known constants set by a design, and y is sampled conditionally on the values of X as in an experiment. For practical purposes, this distinction is often unimportant, since estimation and inference is carried out while conditioning on X. All results stated in this article are within the random design framework.

Classical Linear Regression Model

The classical model focuses on the "finite sample" estimation and inference, meaning that the number of observations n is fixed. This contrasts with the other approaches, which study the asymptotic behaviour of OLS, and in which the number of observations is allowed to grow to infinity.

- Correct specification. The linear functional form is correctly specified.
- Strict exogeneity. The errors in the regression should have conditional mean zero:

$$E[\varepsilon \mid X] = 0.$$

The immediate consequence of the exogeneity assumption is that the errors have mean zero: $E[\varepsilon] = 0$, and that the regressors are uncorrelated with the errors: $E[X2\ \varepsilon] = 0$.

The exogeneity assumption is critical for the OLS theory. If it holds then the regressor variables are called *exogenous*. If it doesn't, then those regressors that are correlated with the error term are called *endogenous*, and then the OLS estimates become invalid. In such case the method of instrumental variables may be used to carry out inference.

- No linear dependence. The regressors in X must all be linearly independent. Mathematically it means that the matrix X must have full column rank almost surely:

  $\Pr[\operatorname{rank}(X) = p] = 1$

  Usually, it is also assumed that the regressors have finite moments up to at least second. In such case the matrix $Q_{xx} = \mathrm{E}[X'X/n]]$ will be finite and positive semi-definite.

  When this assumption is violated the regressors are called linearly dependent or perfectly multicollinear. In such case the value of the regression coefficient β cannot be learned, although prediction of y values is still possible for new values of the regressors that lie in the same linearly dependent subspace.
- Spherical errors:

  $\operatorname{Var}[\varepsilon \mid X] = \sigma^2 I_n,$

  where I_n is an $n \times n$ identity matrix, and σ^2 is a parameter which determines the variance of each observation. This σ^2 is considered a nuisance parameter in the model, although usually it is also estimated. If this assumption is violated then the OLS estimates are still valid, but no longer efficient.

It is customary to split this assumption into two parts:

- o Homoscedasticity: $\mathrm{E}[e_i^2 \mid X] = s^2$, which means that the error term has the same variance s^2 in each observation. When this requirement is violated this is called heteroscedasticity, in such case a more efficient estimator would be weighted least squares. If the errors have infinite variance then the OLS estimates will also have infinite variance (although by the law of large numbers they will nonetheless tend toward the true values so long as the errors have zero mean). In this case, robust estimation techniques are recommended.
- o Nonautocorrelation: the errors are uncorrelated between observations: $\mathrm{E}[\varepsilon_i \varepsilon_j \mid X] = 0$ for $i \neq j$. This assumption may

be violated in the context of time series data, panel data, cluster samples, hierarchical data, repeated measures data, longitudinal data, and other data with dependencies. In such cases generalized least squares provides a better alternative than the OLS.

- Normality. It is sometimes additionally assumed that the errors have normal distribution conditional on the regressors:

  $\varepsilon | X \sim \mathcal{N}(0, \sigma^2 I_n).$

  This assumption is not needed for the validity of the OLS method, although certain additional finite-sample properties can be established in case when it does (especially in the area of hypotheses testing). Also when the errors are normal, the OLS estimator is equivalent to the maximum likelihood estimator (MLE) , and therefore it is asymptotically efficient in the class of all regular estimators.

Independent and Identically Distributed

In some applications, especially with cross-sectional data, an additional assumption is imposed — that all observations are independent and identically distributed (iid). This means that all observations are taken from a random sample which makes all the assumptions listed earlier simpler and easier to interpret. Also this framework allows one to state asymptotic results (as the sample size $n \to \infty$), which are understood as a theoretical possibility of fetching new independent observations from the data generating process. The list of assumptions in this case is:

- iid observations: (x_i, y_i) is independent from, and has the same distribution as, (x_j, y_j) for all $i \neq j$;
- no perfect multicollinearity: $Q_{xx} = \mathrm{E}[x_i x2_i]$ is a positive-definite matrix;
- exogeneity: $\mathrm{E}[\varepsilon_i | x_i] = 0$;
- homoscedasticity: $\mathrm{Var}[\varepsilon_i | x_i] = \sigma^2$.

Time Series Model

- The stochastic process $\{x_i, y_i\}$ is stationary and ergodic;
- The regressors are *predetermined*: $\mathrm{E}[x_i \varepsilon_i] = 0$ for all $i = 1, \ldots, n$;
- The $p \times p$ matrix $Q_{xx} = \mathrm{E}[\ x_i x2\ _i]$ is of full rank, and hence positive-definite;

- $\{x_i\varepsilon_i\}$ is a martingale difference sequence, with a finite matrix of second moments $Q_{xx\varepsilon^2} = \mathrm{E}[\varepsilon_i^2 x_i x2\ _i]$.

Estimation

Suppose b is a "candidate" value for the parameter β. The quantity $y_i - x_i2\ b$ is called the residual for the i-th observation, it measures the vertical distance between the data point (x_i, y_i) and the hyperplane $y = x2\ b$, and thus assesses the degree of fit between the actual data and the model. The sum of squared residuals (SSR) (also called the error sum of squares (ESS) or residual sum of squares (RSS)) is a measure of the overall model fit:

$$S(b) = \sum_{i=1}^{n} (y_i - x_i'b)^2 = (y - Xb)^T (y - Xb),$$

where T denotes the matrix transpose. The value of b which minimizes this sum is called the OLS estimator for β. The function $S(b)$ is quadratic in b with positive-definite Hessian, and therefore this function possesses a unique global minimum at $b = \hat{\beta}$, which can be given by the explicit formula:

$$\hat{\beta} = \arg\min_{b \in \mathbb{R}^p} S(b) = \Big(\frac{1}{n}\sum_{i=1}^{n} x_i x_i'\Big)^{-1} \cdot \frac{1}{n}\sum_{i=1}^{n} x_i y_i$$

or equivalently in matrix form,

$$\hat{\beta} = (X^T X)^{-1} X^T y\,.$$

After we have estimated β, the fitted values (or predicted values) from the regression will be

$$\hat{y} = X\hat{\beta} = Py,$$

where $P = X(X^T X)^{-1} X^T$ is the projection matrix onto the space spanned by the columns of X. This matrix P is also sometimes called the hat matrix because it "puts a hat" onto the variable y. Another matrix, closely related to P is the *annihilator* matrix $M = I_n - P$, this is a projection matrix onto the space orthogonal to X. Both matrices P and M are symmetric and idempotent (meaning that $P^2 = P$), and relate to the data matrix X via identities $PX = X$ and $MX = 0$. Matrix M creates the residuals from the regression:

$$\hat{\varepsilon} = y - X\hat{\beta} = My = M\varepsilon.$$

Using these residuals we can estimate the value of σ^2:

$$s^2 = \frac{\hat{\varepsilon}'\hat{\varepsilon}}{n-p} = \frac{y'My}{n-p} = \frac{S(\hat{\beta})}{n-p}, \qquad \hat{\sigma}^2 = \frac{n-p}{n} s^2$$

The numerator, *n-p*, is the statistical degrees of freedom. The first quantity, s^2, is the OLS estimate for σ^2, whereas the second, σ^2, is the MLE estimate for σ^2. The two estimators are quite similar in large samples; the first one is always unbiased, while the second is biased but minimizes the mean squared error of the estimator. In practice σ^2 is used more often, since it is more convenient for the hypothesis testing. The square root of σ^2 is called the standard error of the regression (SER), or standard error of the equation (SEE).

It is common to assess the goodness-of-fit of the OLS regression by comparing how much the initial variation in the sample can be reduced by regressing onto *X*. The coefficient of determination R^2 is defined as a ratio of "explained" variance to the "total" variance of the dependent variable *y*:

$$R^2 = \frac{\sum(\hat{y}_i - \overline{y})^2}{\sum(y_i - \overline{y})^2} = \frac{y'P'LPy}{y'Ly} = 1 - \frac{y'My}{y'Ly} = 1 - \frac{\text{SSR}}{\text{TSS}}$$

where TSS is the total sum of squares for the dependent variable, $L = I_n - 112\,/n$, and 1 is an $n\times1$ vector of ones. (*L* is a "centering matrix" which is equivalent to regression on a constant; it simply subtracts the mean from a variable.) In order for R^2 to be meaningful, the matrix *X* of data on regressors must contain a column vector of ones to represent the constant whose coefficient is the regression intercept. In that case, R^2 will always be a number between 0 and 1, with values close to 1 indicating a good degree of fit.

Simple Regression Model

If the data matrix *X* contains only two variables: a constant, and a scalar regressor x_i, then this is called the "simple regression model". This case is often considered in the beginner statistics classes, as it provides much simpler formulas even suitable for manual calculation. The vectors of parameters in such model is 2-dimensional, and is commonly denoted as (α, β):

$$y_i = \alpha + \beta x_i + \varepsilon_i.$$

The least squares estimates in this case are given by simple formulas

$$\hat{\beta} = \frac{\sum x_i y_i - \frac{1}{n}\sum x_i \sum y_i}{\sum x_i^2 - \frac{1}{n}(\sum x_i)^2} = \frac{\text{Cov}[x,y]}{\text{Var}[x]}, \quad \hat{\alpha} = \overline{y} - \hat{\beta}\overline{x}.$$

Alternative derivations

In the previous section the least squares estimator $\hat{\beta}$ was obtained as a value that minimizes the sum of squared residuals of the model. However it is also possible to derive the same estimator from other approaches. In all cases the formula for OLS estimator remains the same: $\beta = (X'X)^{-1}X'y$, the only difference is in how we interpret this result.

Geometric Approach

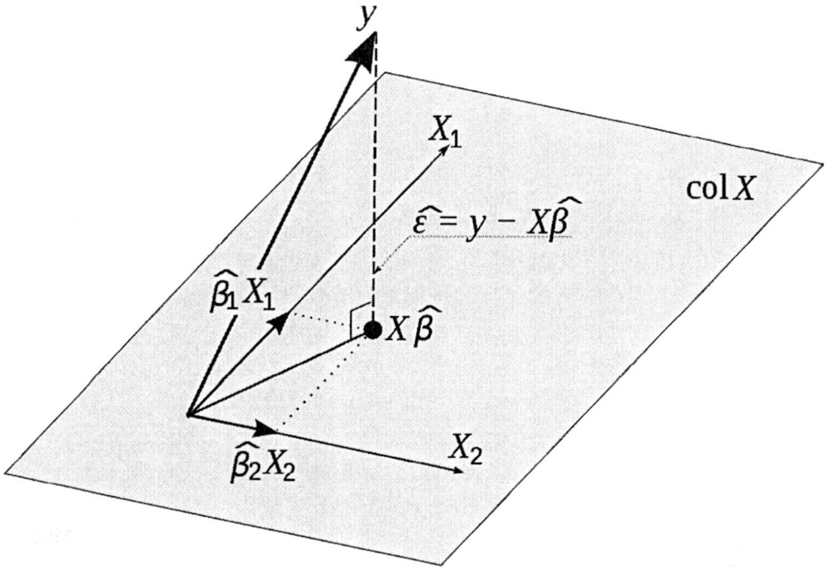

Figure: *OLS estimation can be viewed as a projection onto the linear space spanned by the regressors.*

For mathematicians, OLS is an approximate solution to an overdetermined system of linear equations $X\beta \approx y$, where β is the unknown. Assuming the system cannot be solved exactly (the number of equations n is much larger than the number of unknowns p), we are looking for a solution that could provide the smallest discrepancy between the right- and left- hand sides. In other words, we are looking for the solution that satisfies

$$\hat{\beta} = \arg\min_{\beta} \| y - X\beta \|,$$

where $\|\cdot\|$ is the standard L^2 norm in the n-dimensional Euclidean space R^n. The predicted quantity $X\beta$ is just a certain linear combination of the vectors of regressors. Thus, the residual vector $y - X\beta$ will have the smallest length when y is projected orthogonally onto the linear subspace spanned by the columns of X. The OLS estimator $\hat{\beta}$ in this case can be interpreted as the coefficients of vector decomposition of $\hat{y} = py$ along the basis of X.

Maximum Likelihood

The OLS estimator is identical to the maximum likelihood estimator (MLE) under the normality assumption for the error terms. This normality assumption has historical importance, as it provided the basis for the early work in linear regression analysis by Yule and Pearson. From the properties of MLE, we can infer that the OLS estimator is asymptotically efficient (in the sense of attaining the Cramér-Rao bound for variance) if the normality assumption is satisfied.

Generalized Method of Moments

In iid case the OLS estimator can also be viewed as a GMM estimator arising from the moment conditions

$$\mathrm{E}\big[x_i(y_i - x_{i'}\beta)\big] = 0.$$

These moment conditions state that the regressors should be uncorrelated with the errors. Since x_i is a p-vector, the number of moment conditions is equal to the dimension of the parameter vector β, and thus the system is exactly identified. This is the so-called classical GMM case, when the estimator does not depend on the choice of the weighting matrix.

Note that the original strict exogeneity assumption $\mathrm{E}[\varepsilon_i \mid x_i] = 0$ implies a far richer set of moment conditions than stated above. In particular, this assumption implies that for any vector-function f, the moment condition $\mathrm{E}[f(x_i) \cdot \varepsilon_i] = 0$ will hold. However it can be shown using the Gauss–Markov theorem that the optimal choice of function f is to take $f(x) = x$, which results in the moment equation posted above.

Finite Sample Properties

First of all, under the *strict exogeneity* assumption the OLS estimators $\hat{\beta}$ and s^2 are unbiased, meaning that their expected values coincide with the true values of the parameters:

$$\mathrm{E}[\hat{\beta} \mid X] = \beta, \quad \mathrm{E}[s^2 \mid X] = \sigma^2.$$

If the strict exogeneity does not hold (as is the case with many time series models, where exogeneity is assumed only with respect to the past shocks but not the future ones), then these estimators will be biased in finite samples.

The variance-covariance matrix of $\hat{\beta}$ is equal to

$$\mathrm{Var}[\hat{\beta}|\,X] = \sigma^2 (X'X)^{-1}.$$

In particular, the standard error of each coefficient $\hat{\beta}_j$ is equal to square root of the j-th diagonal element of this matrix. The estimate of this standard error is obtained by replacing the unknown quantity σ^2 with its estimate s^2. Thus,

$$\widehat{\mathrm{s.e}}(\hat{\beta}_j) = \sqrt{s^2 (X'X)^{-1}_{jj}}$$

It can also be easily shown that the estimator is uncorrelated with the residuals from the model:

The Gauss–Markov theorem states that under the *spherical errors* assumption (that is, the errors should be uncorrelated and homoscedastic) the estimator $\hat{\beta}$ is efficient in the class of linear unbiased estimators. This is called the best linear unbiased estimator (BLUE). Efficiency should be understood as if we were to find some other estimator $\tilde{\beta}$ which would be linear in y and unbiased, then

$$\mathrm{Var}[\tilde{\beta}|\,X] - \mathrm{Var}[\hat{\beta}|\,X] \geq 0$$

in the sense that this is a nonnegative-definite matrix. This theorem establishes optimality only in the class of linear unbiased estimators, which is quite restrictive. Depending on the distribution of the error terms ε, other, non-linear estimators may provide better results than OLS.

Assuming Normality

The properties listed so far are all valid regardless of the underlying distribution of the error terms. However if you are willing to assume that the *normality assumption* holds (that is, that $\varepsilon \sim N(0, \sigma^2 I_n)$), then additional properties of the OLS estimators can be stated.

The estimator is normally distributed, with mean and variance as given before:

$$\hat{\beta} \sim \mathcal{N}\big(\beta, \sigma^2 (X'X)^{-1}\big)$$

This estimator reaches the Cramér–Rao bound for the model, and thus is optimal in the class of all unbiased estimators. Note that unlike the Gauss–Markov theorem, this result establishes optimality among both linear and non-linear estimators, but only in the case of normally distributed error terms.

The estimator s^2 will be proportional to the chi-squared distribution:

$$s^2 \sim \frac{\sigma^2}{n-p} \cdot \chi^2_{n-p}$$

The variance of this estimator is equal to $2\sigma^4/(n-p)$, which does not attain the Cramér–Rao bound of $2\sigma^4/n$. However it was shown that there are no unbiased estimators of σ^2 with variance smaller than that of the estimator s^2. If we are willing to allow biased estimators, and consider the class of estimators that are proportional to the sum of squared residuals (SSR) of the model, then the best (in the sense of the mean squared error) estimator in this class will be $\tilde{\sigma}^2 = \text{SSR}/(n-p+2)$, which even beats the Cramér–Rao bound in case when there is only one regressor (p = 1).

Moreover, the estimators $\hat{\beta}$ and s^2 are independent, the fact which comes in useful when constructing the t- and F-tests for the regression.

Influential Observations

As was mentioned before, the estimator $\hat{\beta}$ is linear in y, meaning that it represents a linear combination of the dependent variables y_i's. The weights in this linear combination are functions of the regressors X, and generally are unequal. The observations with high weights are called influential because they have a more pronounced effect on the value of the estimator.

To analyze which observations are influential we remove a specific j-th observation and consider how much the estimated quantities are going to change (similarly to the jackknife method). It can be shown that the change in the OLS estimator for β will be equal to

$$\hat{\beta}^{(j)} - \hat{\beta} = -\frac{1}{1-h_j}(X'X)^{-1}x_j'\hat{\varepsilon}_j,$$

where $h_j = x_j2\,(X2\,X)^{-1}x_j$ is the j-th diagonal element of the hat matrix P, and x_j is the vector of regressors corresponding to the j-th observation. Similarly, the change in the predicted value for j-th observation resulting from omitting that observation from the dataset will be equal to

$$\hat{y}_j^{(j)} - \hat{y}_j = x_j'\hat{\beta}^{(j)} - x_j'\hat{\beta} = -\frac{h_j}{1-h_j}\hat{\varepsilon}_j$$

From the properties of the hat matrix, $0 \le h_j \le 1$, and they sum up to p, so that on average $h_j \approx p/n$. These quantities h_j are called the leverages', *and observations with high* h_js — leverage points. Usually the observations with high leverage ought to be scrutinized more carefully, in case they are erroneous, or outliers, or in some other way atypical of the rest of the dataset.

Partitioned Regression

Sometimes the variables and corresponding parameters in the regression can be logically split into two groups, so that the regression takes form

$$y = X_1\beta_1 + X_2\beta_2 + \varepsilon,$$

where X_1 and X_2 have dimensions $n \times p_1$, $n \times p_2$, and β_1, β_2 are $p_1 \times 1$ and $p_2 \times 1$ vectors, with $p_1 + p_2 = p$.

The Frisch–Waugh–Lovell theorem states that in this regression the residuals $\hat{\varepsilon}$ and the OLS estimate $\hat{\beta}_2$ will be numerically identical to the residuals and the OLS estimate for β_2 in the following regression:

$$M_1 y = M_1 X_2 \beta_2 + \eta,$$

where M_1 is the annihilator matrix for regressors X_1.

The theorem can be used to establish a number of theoretical results. For example, having a regression with a constant and another regressor is equivalent to subtracting the means from the dependent variable and the regressor and then running the regression for the demeaned variables but without the constant term.

Constrained Estimation

Suppose it is known that the coefficients in the regression satisfy a system of linear equations

$$H_0\colon\ Q'\beta = c,$$

where Q is a $p \times q$ matrix of full rank, and c is a $q \times 1$ vector of known constants, where $q<p$. In this case least squares estimation is equivalent to minimizing the sum of squared residuals of the model subject to the constraint H_0. The constrained least squares (CLS) estimator can be given by an explicit formula:

$$\hat{\beta}^c = \hat{\beta} - (X'X)^{-1}Q\left(Q'(X'X)^{-1}Q\right)^{-1}(Q'\hat{\beta} - c)$$

This expression for the constrained estimator is valid as long as the matrix *X2 X* is invertible. It was assumed from the beginning of this article that this matrix is of full rank, and it was noted that when the rank condition fails, β will not be identifiable. However it may happen that adding the restriction H_0 makes β identifiable, in which case one would like to find the formula for the estimator. The estimator is equal to

$$\hat{\beta}^c = R(R'X'XR)^{-1}R'X'y + \left(I_p - R(R'X'XR)^{-1}R'X'X\right)Q(Q'Q)^{-1}c,$$

where R is a $p \times (p-q)$ matrix such that the matrix $[Q\ R]$ is non-singular, and *R2 Q* = 0. Such a matrix can always be found, although generally it is not unique. The second formula coincides with the first in case when *X2 X* is invertible.

Large Sample Properties

The least squares estimators are point estimates of the linear regression model parameters β. However generally we also want to know how close those estimates might be to the true values of parameters. In other words, we want to construct the interval estimates.

Since we haven't made any assumption about the distribution of error term ε_i, it is impossible to infer the distribution of the estimators $\hat{\beta}$ and $\hat{\sigma}^2$. Nevertheless, we can apply the law of large numbers and central limit theorem to derive their *asymptotic* properties as sample size n goes to infinity. While the sample size is necessarily finite, it is customary to assume that n is "large enough" so that the true distribution of the OLS estimator is close to its asymptotic limit, and the former may be approximately replaced by the latter.

We can show that under the model assumptions, least squares estimator for β is consistent (that is $\hat{\beta}$ converges in probability to β) and asymptotically normal:

$$\sqrt{n}(\hat{\beta} - \beta) \xrightarrow{d} \mathcal{N}\left(0, \sigma^2 Q_{xx}^{-1}\right),$$

where $Q_{xx} = X'X.$

Using this asymptotic distribution, approximate two-sided confidence intervals for the j-th component of the vector $\hat{\beta}$ can be constructed as

$$\beta_j \in \left[\, \hat{\beta}_j \pm q_{1-\alpha/2}^{\mathcal{N}(0,1)} \sqrt{\tfrac{1}{n}\hat{\sigma}^2\left[Q_{xx}^{-1}\right]_{jj}}\, \right] \text{ at the } 1-\alpha \text{ confidence level,}$$

where q denotes the quantile function of standard normal distribution, and $[\cdot]_{jj}$ is the j-th diagonal element of a matrix.

Similarly, the least squares estimator for σ^2 is also consistent and asymptotically normal (provided that the fourth moment of ε_i exists) with limiting distribution

$$\sqrt{n}(\hat{\sigma}^2 - \sigma^2) \xrightarrow{d} \mathcal{N}\left(0, \mathrm{E}[\varepsilon_i^4] - \sigma^4\right).$$

These asymptotic distributions can be used for prediction, testing hypotheses, constructing other estimators, etc.. As an example consider the problem of prediction. Suppose x_0 is some point within the domain of distribution of the regressors, and one wants to know what the response variable would have been at that point. The mean response is the quantity $y_0 = x_0'\beta$, whereas the predicted response is $\hat{y}_0 = x_0'\hat{\beta}$. . Clearly the predicted response is a random variable, its distribution can be derived from that of $\hat{\beta}$:

$$\sqrt{n}(\hat{y}_0 - y_0) \xrightarrow{d} \mathcal{N}\left(0, \sigma^2 x_0' Q_{xx}^{-1} x_0\right),$$

which allows construct confidence intervals for mean response y_0 to be constructed:

$$y_0 \in \left[\, x_0'\hat{\beta} \pm q_{1-\alpha/2}^{\mathcal{N}(0,1)} \sqrt{\tfrac{1}{n}\hat{\sigma}^2 x_0' Q_{xx}^{-1} x_0}\, \right] \text{ at the } 1-\alpha \text{ confidence level.}$$

Example with Real Data

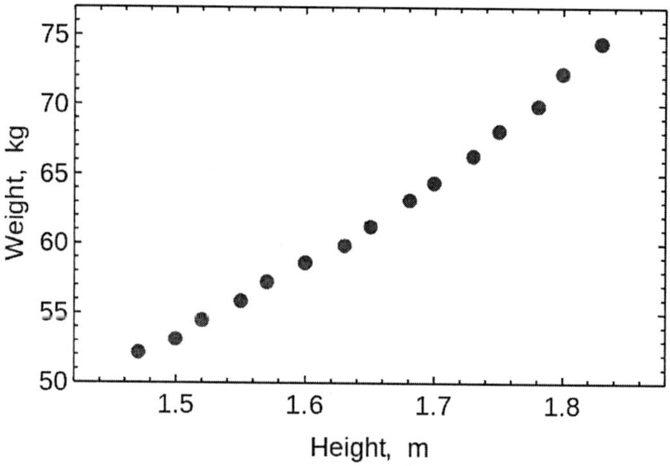

Figure: *Scatterplot of the data; the relationship is slightly curved but close to linear.*

When only one dependent variable is being modelled, a scatterplot will suggest the form and strength of the relationship between the dependent variable and regressors. It might also reveal outliers, heteroscedasticity, and other aspects of the data that may complicate the interpretation of a fitted regression model. The scatterplot suggests that the relationship is strong and can be approximated as a quadratic function. OLS can handle non-linear relationships by introducing the regressor HEIGHT². The regression model then becomes a multiple linear model:

$$w_i = \beta_1 + \beta_2 h_i + \beta_3 h_i^2 + \varepsilon_i.$$

The output from most popular statistical packages will look similar to this:

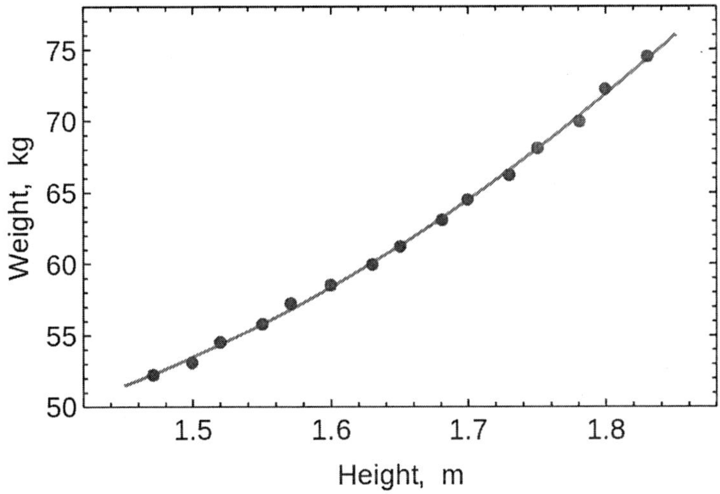

Figure: *Fitted regression*

Method: Least Squares Dependent variable: WEIGHT Included observations: 15					
Variable	**Coefficient**		**Std.Error**	**t-statistic**	**p-value**
β	128.8128		16.3083	7.8986	0.0000
h	−143.1620		19.8332	−7.2183	0.0000
h^2	61.9603		6.0084	10.3122	0.0000
R^2	0.9989		S.E. of regression		0.2516
Adjusted R^2	0.9987		Model sum-of-sq		692.61
Log-likelihood	1.0890		Residual sum-of-sq		0.7595
Durbin–Watson stats.	2.1013		Total sum-of-sq		693.37
Akaike criterion	0.2548		F-statistic		5471.2
Schwarz criterion	0.3964		p-value (F-stat)		0.0000

In this table:

- The *Coefficient* column gives the least squares estimates of parameters β_j

- The *Std. errors* column shows standard errors of each coefficient estimate: $\hat{\sigma}_j = \left(\frac{1}{n}\hat{\sigma}^2[\hat{Q}_{xx}^{-1}]_{jj}\right)^{1/2}$
- The *t-statistic* and *p-value* columns are testing whether any of the coefficients might be equal to zero. The *t*-statistic is calculated simply as $t = \hat{\beta}_j / \hat{\sigma}_j$. If the errors ε approximately follow a normal distribution, *t* follows a Student-t distribution. Under weaker conditions, *t* is asymptotically normal. Large values of *t* indicate that the null hypothesis can be rejected and that the corresponding coefficient is not zero. The second column, *p*-value, expresses the results of the hypothesis test as a significance level. Conventionally, *p*-values smaller than 0.05 are taken as evidence that the population coefficient is nonzero.
- *R-squared* is the coefficient of determination indicating goodness-of-fit of the regression. This statistic will be equal to one if fit is perfect, and to zero when regressors *X* have no explanatory power whatsoever. This is a biased estimate of the population *R-squared*, and will never decrease if additional regressors are added, even if they are irrelevant.
- *Adjusted R-squared* is a slightly modified version of R^2, designed to penalize for the excess number of regressors which do not add to the explanatory power of the regression. This statistic is always smaller than R^2, can decrease as new regressors are added, and even be negative for poorly fitting models:

  $$\bar{R}^2 = 1 - \tfrac{n-1}{n-p}(1 - R^2)$$

- *Log-likelihood* is calculated under the assumption that errors follow normal distribution. Even though the assumption is not very reasonable, this statistic may still find its use in conducting LR tests.
- *Durbin–Watson statistic* tests whether there is any evidence of serial correlation between the residuals. As a rule of thumb, the value smaller than 2 will be an evidence of positive correlation.
- *Akaike information criterion* and *Schwarz criterion* are both used for model selection. Generally when comparing two alternative models, smaller values of one of these criteria will indicate a better model.

- *Standard error of regression* is an estimate of σ, standard error of the error term.
- *Total sum of squares*, *model sum of squared*, and *residual sum of squares* tell us how much of the initial variation in the sample were explained by the regression.
- *F-statistic* tries to test the hypothesis that all coefficients (except the intercept) are equal to zero. This statistic has $F(p1,n-p)$ distribution under the null hypothesis and normality assumption, and its *p-value* indicates probability that the hypothesis is indeed true. Note that when errors are not normal this statistic becomes invalid, and other tests such as for example Wald test or LR test should be used.

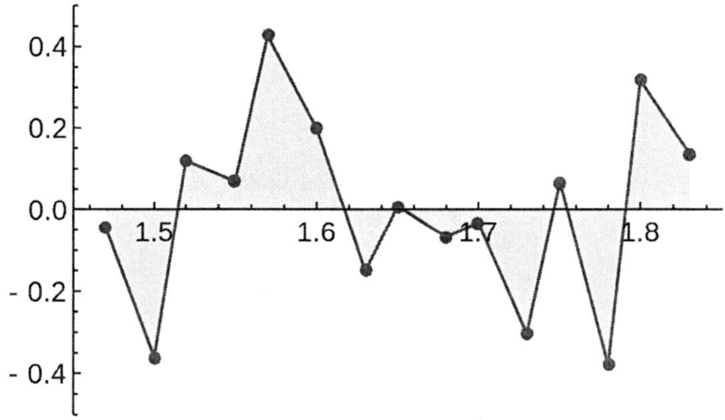

Figure: *Residuals plot*

Ordinary least squares analysis often includes the use of diagnostic plots designed to detect departures of the data from the assumed form of the model. These are some of the common diagnostic plots:

- Residuals against the explanatory variables in the model. A non-linear relation between these variables suggests that the linearity of the conditional mean function may not hold. Different levels of variability in the residuals for different levels of the explanatory variables suggests possible heteroscedasticity.
- Residuals against explanatory variables not in the model. Any relation of the residuals to these variables would suggest considering these variables for inclusion in the model.
- Residuals against the fitted values, $\hat{y}$.
- Residuals against the preceding residual. This plot may identify serial correlations in the residuals.

An important consideration when carrying out statistical inference using regression models is how the data were sampled. In this example, the data are averages rather than measurements on individual women. The fit of the model is very good, but this does not imply that the weight of an individual woman can be predicted with high accuracy based only on her height.

Sensitivity to Rounding

This example also demonstrates that coefficients determined by these calculations are sensitive to how the data is prepared. The heights were originally given rounded to the nearest inch and have been converted and rounded to the nearest centimetre. Since the conversion factor is one inch to 2.54cm this is *not* an exact conversion. The original inches can be recovered by Round(x/0.0254) and then re-converted to metric without rounding. If this is done the results become:

const	height	Height^2	
128.8128	-143.162	61.96033	converted to metric with rounding.
119.0205	-131.5076	58.5046	converted to metric without rounding.

Using either of these equations to predict the weight of a 5' 6" (1.6764m) woman gives similar values: 62.94kg with rounding vs. 62.98kg without rounding. Thus a seemingly small variation in the data has a real effect on the coefficients but a small effect on the results of the equation.

Bias of an Estimator

In statistics, the bias (or bias function) of an estimator is the difference between this estimator's expected value and the true value of the parameter being estimated. An estimator or decision rule with zero bias is called unbiased. Otherwise the estimator is said to be biased.

In ordinary English, the word *bias* is pejorative. In statistics, there are problems for which it may be good to use an estimator with a small, but nonzero, bias. In some cases, an estimator with a small bias may have lesser mean squared error or be *median*-unbiased (rather than *mean*-unbiased, the standard unbiasedness property). The property of *median*-unbiasedness is invariant under monotone transformations, while the property of *mean*-unbiasedness may be lost under nonlinear transformations.

Definition

Suppose we have a statistical model parameterized by θ giving rise to a probability distribution for observed data, $P(x \mid \theta)$, and a statistic $\hat{\theta}$ which serves as an estimator of θ based on any observed data x. That is, we assume that our data follow some unknown distribution $P(x \mid \theta)$ (where θ is a fixed constant that is part of this distribution, but is unknown), and then we construct some estimator that maps observed data to values that we hope are close to θ. Then the bias of this estimator is defined to be

$$\operatorname{Bias}[\hat{\theta}] = \operatorname{E}[\hat{\theta}] - \theta = \operatorname{E}[\hat{\theta} - \theta],$$

where E[ ] denotes expected value over the distribution $P(x \mid \theta)$, i.e. averaging over all possible observations x.

An estimator is said to be unbiased if its bias is equal to zero for all values of parameter θ.

There are more general notions of bias and unbiasedness. What this article calls "bias" is called "*mean*-bias", to distinguish *mean*-bias from the other notions, with the notable ones being "*median*-unbiased" estimators.

In a simulation experiment concerning the properties of an estimator, the bias of the estimator may be assessed using the mean signed difference.

Examples

Sample Variance: Suppose $X_1, \ldots, X_n$ are independent and identically distributed (i.i.d.) random variables with expectation μ and variance σ^2. If the sample mean and uncorrected sample variance are defined as

$$\bar{X} = \frac{1}{n}\sum_{i=1}^{n} X_i, \qquad S^2 = \frac{1}{n}\sum_{i=1}^{n}\left(X_i - \bar{X}\right)^2,$$

then S^2 is a biased estimator of σ^2, because

$$\operatorname{E}[S^2] = \operatorname{E}\left[\frac{1}{n}\sum_{i=1}^{n}\left(X_i - \bar{X}\right)^2\right] = \operatorname{E}\big[\frac{1}{n}\sum_{i=1}^{n}\big((X_i - \mu) - (\bar{X} - \mu)\big)^2\big]$$

$$[8pt] = \operatorname{E}\big[\frac{1}{n}\sum_{i=1}^{n}(X_i - \mu)^2 - 2(\bar{X} - \mu)\frac{1}{n}\sum_{i=1}^{n}(X_i - \mu) + (\bar{X} - \mu)^2\big]$$

$$[8pt] = \operatorname{E}\big[\frac{1}{n}\sum_{i=1}^{n}(X_i - \mu)^2 - (\bar{X} - \mu)^2\big] = \sigma^2 - \operatorname{E}\left[(\bar{X} - \mu)^2\right] < \sigma^2.$$

In other words, the expected value of the uncorrected sample variance does not equal the population variance σ^2, unless multiplied by a normalization factor. The sample mean, on the other hand, is an unbiased estimator of the population mean μ.

The reason that S^2 is biased stems from the fact that the sample mean is an ordinary least squares (OLS) estimator for μ: $\bar{X}$ is the number that makes the sum $\sum_{i=1}^{N}(X_i - \bar{X})^2$ as small as possible. That is, when any other number is plugged into this sum, the sum can only increase. In particular, the choice $\mu \neq \bar{X}$ gives,

$$\frac{1}{n}\sum_{i=1}^{n}(X_i - \bar{X})^2 < \frac{1}{n}\sum_{i=1}^{n}(X_i - \mu)^2,$$

and then

$$\mathrm{E}[S^2] = \mathrm{E}\Big[\frac{1}{n}\sum_{i=1}^{n}(X_i - \bar{X})^2\Big] < \mathrm{E}\Big[\frac{1}{n}\sum_{i=1}^{n}(X_i - \mu)^2\Big] = \sigma^2.$$

Note that the usual definition of sample variance is

$$s^2 = \frac{1}{n-1}\sum_{i=1}^{n}(X_i - \bar{X})^2,$$

and this is an unbiased estimator of the population variance. This can be seen by noting the following formula, which follows from the Bienaymé formula, for the term in the inequality for the expectation of the uncorrected sample variance above:

$$\mathrm{E}\big[(\bar{X} - \mu)^2\big] = \frac{1}{n}\sigma^2.$$

The ratio between the biased (uncorrected) and unbiased estimates of the variance is known as Bessel's correction.

Estimating a Poisson Probability

A far more extreme case of a biased estimator being better than any unbiased estimator arises from the Poisson distribution. Suppose that X has a Poisson distribution with expectation λ. Suppose it is desired to estimate

$$\mathrm{P}(X = 0)^2 = e^{-2\lambda}.$$

(For example, when incoming calls at a telephone switchboard are modelled as a Poisson process, and λ is the average number of calls per minute, then $e^{-2\lambda}$ is the probability that no calls arrive in the next two minutes.)

Since the expectation of an unbiased estimator δ(X) is equal to the estimand, i.e.

$$E(\delta(X)) = \sum_{x=0}^{\infty} \delta(x) \frac{\lambda^x e^{-\lambda}}{x!} = e^{-2\lambda},$$

the only function of the data constituting an unbiased estimator is

$$\delta(x) = (-1)^x.$$

To see this, note that when decomposing $e^{-\lambda}$ from the above expression for expectation, the sum that is left is a Taylor series expansion of $e^{-\lambda}$ as well, yielding $e^{-\lambda}e^{-\lambda} = e^{-2\lambda}$.

If the observed value of X is 100, then the estimate is 1, although the true value of the quantity being estimated is very likely to be near 0, which is the opposite extreme. And, if X is observed to be 101, then the estimate is even more absurd: It is –1, although the quantity being estimated must be positive.

The (biased) maximum likelihood estimator

$$e^{-2\bar{X}}$$

is far better than this unbiased estimator. Not only is its value always positive but it is also more accurate in the sense that its mean squared error

$$e^{-4\lambda} - 2e^{\lambda(1/e^2-3)} + e^{\lambda(1/e^4-1)}$$

is smaller; compare the unbiased estimator's MSE of

$$1 - e^{-4\lambda}.$$

The MSEs are functions of the true value λ. The bias of the maximum-likelihood estimator is:

$$e^{-2\lambda} - e^{\lambda(1/e^2-1)}.$$

Maximum of a Discrete Uniform Distribution

The bias of maximum-likelihood estimators can be substantial. Consider a case where n tickets numbered from 1 through to n are placed in a box and one is selected at random, giving a value X.

If n is unknown, then the maximum-likelihood estimator of n is X, even though the expectation of X is only $(n + 1)/2$; we can be certain only that n is at least X and is probably more. In this case, the natural unbiased estimator is $2X - 1$.

Median-unbiased Estimators

The theory of median-unbiased estimators was revived by George W. Brown in 1947:

An estimate of a one-dimensional parameter θ will be said to be median-unbiased, if, for fixed θ, the median of the distribution of the estimate is at the value θ; i.e., the estimate underestimates just as often as it overestimates. This requirement seems for most purposes to accomplish as much as the mean-unbiased requirement and has the additional property that it is invariant under one-to-one transformation.

Further properties of median-unbiased estimators have been noted by Lehmann, Birnbaum, van der Vaart and Pfanzagl. In particular, median-unbiased estimators exist in cases where mean-unbiased and maximum-likelihood estimators do not exist. Besides being invariant under one-to-one transformations, median-unbiased estimators have surprising robustness.

Bias with Respect to Other Loss Functions

Any *mean*-unbiased minimum-variance estimator minimizes the risk (expected loss) with respect to the squared-error loss function, as observed by Gauss. A *median*-unbiased estimator minimizes the risk with respect to the absolute loss function, as observed by Laplace. Other loss functions are used in statistical theory, particularly in robust statistics.

Effect of Transformations

Note that, when a transformation is applied to a mean-unbiased estimator, the result need not be a mean-unbiased estimator of its corresponding population statistic. By Jensen's inequality, a convex function as transformation will introduce positive bias, while a concave function will introduce negative bias, and a function of mixed convexity may introduce bias in either direction, depending on the specific function and distribution. That is, for a non-linear function f and a mean-unbiased estimator U of a parameter p, the composite estimator $f(U)$ need not be a mean-unbiased estimator of $f(p)$. For example, the square root of the unbiased estimator of the population variance is *not* a mean-unbiased estimator of the population standard deviation: the square root of the unbiased sample variance, the corrected sample standard deviation, is biased. The bias depends both on the sampling distribution of the estimator and on the transform, and can be quite involved to calculate.

Bias, Variance and Mean Squared Error

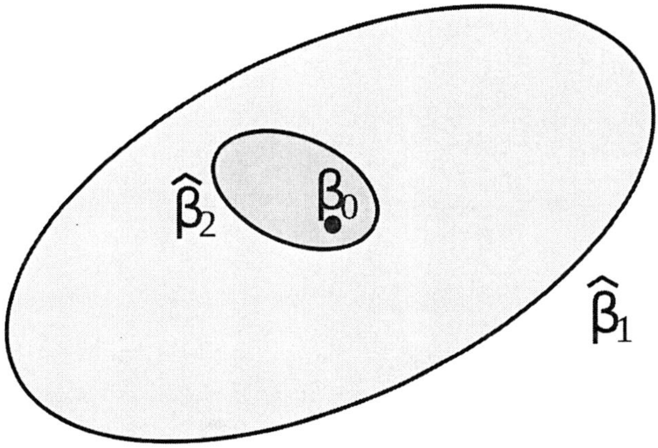

Figure: *Sampling distributions of two alternative estimators for a parameter β_0. Although $\hat{\beta}_1$ is unbiased, it is clearly inferior to the biased $\hat{\beta}_2$.*

Ridge regression is one example of a technique where allowing a little bias may lead to a considerable reduction in variance, and more reliable estimates overall. While bias quantifies the *average* difference to be expected between an estimator and an underlying parameter, an estimator based on a finite sample can additionally be expected to differ from the parameter due to the randomness in the sample.

One measure which is used to try to reflect both types of difference is the expected mean square error,

$$\text{MSE}(\hat{\theta}) = \text{E}\left[(\hat{\theta} - \theta)^2\right].$$

This can be shown to be equal to the square of the expected bias, plus the expected variance:

$$\begin{aligned}\text{MSE}(\hat{\theta}) = &\quad (\text{E}[\hat{\theta} - \theta])^2 + \text{E}[(\hat{\theta} - \text{E}[\hat{\theta}])^2] \\ = &\quad (\text{Bias}(\hat{\theta}, \theta))^2 + \text{Var}(\hat{\theta})\end{aligned}$$

An estimator that minimises the bias will not necessarily minimise the expected mean square error.

Example: Estimation of Population Variance

For example, suppose an estimator of the form

$$T^2 = c\sum_{i=1}^{n}\left(X_i - \bar{X}\right)^2 = cnS^2$$

is sought for the population variance as above, but this time to minimise the MSE:

$$\begin{aligned} \text{MSE} &= \text{E}\left[(T^2 - \sigma^2)^2\right] \\ &= \left(\text{E}\left[T^2 - \sigma^2\right]\right)^2 + \text{Var}(T^2) \end{aligned}$$

If the variables $X_1 \ldots X_n$ follow a normal distribution, then nS^2/σ^2 has a chi-squared distribution with $n - 1$ degrees of freedom, giving:

$$\text{E}[nS^2] = (n-1)\sigma^2 \text{ and } \text{Var}(nS^2) = 2(n-1)\sigma^4.$$

and so

$$\text{MSE} = (c(n-1)-1)^2\sigma^4 + 2c^2(n-1)\sigma^4$$

With a little algebra it can be confirmed that it is $c = 1/(n + 1)$ which minimises this combined loss function, rather than $c = 1/(n - 1)$ which minimises just the bias term. More generally it is only in restricted classes of problems that there will be an estimator that minimises the MSE independently of the parameter values.

However it is very common that there may be perceived to be a *bias–variance tradeoff*, such that a small increase in bias can be traded for a larger decrease in variance, resulting in a more desirable estimator overall.

Bayesian View

Most bayesians are rather unconcerned about unbiasedness (at least in the formal sampling-theory sense above) of their estimates. For example, Gelman *et al* (1995) write: "From a Bayesian perspective, the principle of unbiasedness is reasonable in the limit of large samples, but otherwise it is potentially misleading."

Fundamentally, the difference between the Bayesian approach and the sampling-theory approach above is that in the sampling-theory approach the parameter is taken as fixed, and then probability distributions of a statistic are considered, based on the predicted sampling distribution of the data. For a Bayesian, however, it is the *data* which is known, and fixed, and it is the unknown parameter for which an attempt is made to construct a probability distribution, using Bayes' theorem:

$$p(\theta \mid D, I) \propto p(\theta \mid I) p(D \mid \theta, I)$$

Here the second term, the likelihood of the data given the unknown parameter value θ, depends just on the data obtained and the modelling of the data generation process. However a Bayesian calculation also includes the first term, the prior probability for θ, which takes account of everything the analyst may know or suspect about θ *before* the data comes in. This information plays no part in the sampling-theory approach; indeed any attempt to include it would be considered "bias" away from what was pointed to purely by the data. To the extent that Bayesian calculations include prior information, it is therefore essentially inevitable that their results will not be "unbiased" in sampling theory terms.

But the results of a Bayesian approach can differ from the sampling theory approach even if the Bayesian tries to adopt an "uninformative" prior. For example, consider again the estimation of an unknown population variance σ^2 of a Normal distribution with unknown mean, where it is desired to optimise c in the expected loss function

$$\text{Expected Loss} = \mathrm{E}\left[\left(cnS^2 - \sigma^2\right)^2\right] = \mathrm{E}\left[\sigma^4\left(cn\frac{S^2}{\sigma^2} - 1\right)^2\right]$$

A standard choice of uninformative prior for this problem is the Jeffreys prior, $p(\sigma^2) \propto 1/\sigma^2$, which is equivalent to adopting a rescaling-invariant flat prior for $\ln(\sigma^2)$.

One consequence of adopting this prior is that S^2/σ^2 remains a pivotal quantity, i.e. the probability distribution of S^2/σ^2 depends only on S^2/σ^2, independent of the value of S^2 or σ^2:

$$p\left(\frac{S^2}{\sigma^2} \mid S^2\right) = p\left(\frac{S^2}{\sigma^2} \mid \sigma^2\right) = g\left(\frac{S^2}{\sigma^2}\right)$$

However, whilst

$$\mathrm{E}_{p(S^2|\sigma^2)}\left[\sigma^4\left(cn\frac{S^2}{\sigma^2} - 1\right)^2\right] = \sigma^4\,\mathrm{E}_{p(S^2|\sigma^2)}\left[\left(cn\frac{S^2}{\sigma^2} - 1\right)^2\right]$$

in contrast

$$\mathrm{E}_{p(\sigma^2|S^2)}\left[\sigma^4\left(cn\frac{S^2}{\sigma^2} - 1\right)^2\right] \neq \sigma^4\,\mathrm{E}_{p(\sigma^2|S^2)}\left[\left(cn\frac{S^2}{\sigma^2} - 1\right)^2\right]$$

— when the expectation is taken over the probability distribution of σ^2 given S^2, as it is in the Bayesian case, rather than S^2 given σ^2, one can no longer take σ^4 as a constant and factor it out. The consequence of this is that, compared to the sampling-theory calculation, the Bayesian calculation puts more weight on larger values of σ^2, properly taking into account (as the sampling-theory calculation cannot) that under this squared-loss function the consequence of underestimating large values of σ^2 is more costly in squared-loss terms than that of overestimating small values of σ^2.

The worked-out Bayesian calculation gives a scaled inverse chi-squared distribution with $n - 1$ degrees of freedom for the posterior probability distribution of σ^2. The expected loss is minimised when $cnS^2 = \langle\sigma^2\rangle$; this occurs when $c = 1/(n - 3)$.

Even with an uninformative prior, therefore, a Bayesian calculation may not give the same expected-loss minimising result as the corresponding sampling-theory calculation.

Consistent Estimator

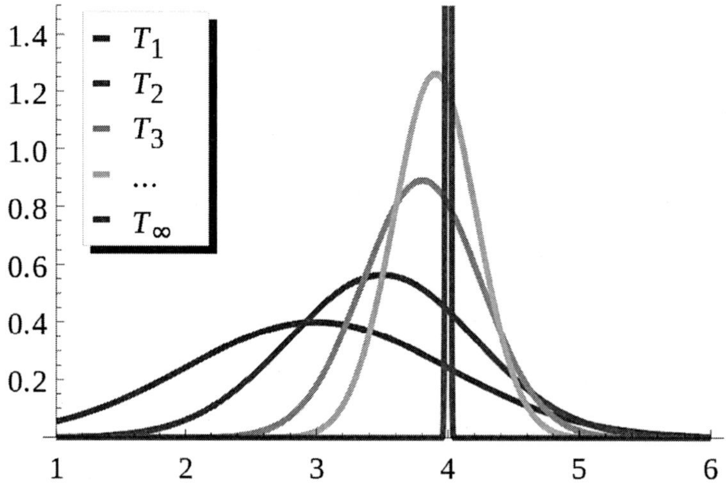

$\{T_1, T_2, T_3, \ldots\}$ is a sequence of estimators for parameter θ_0, the true value of which is 4. This sequence is consistent: the estimators are getting more and more concentrated near the true value θ_0; at the same time, these estimators are biased. The limiting distribution of the sequence is a degenerate random variable which equals θ_0 with probability 1.

In statistics, a consistent estimator or asymptotically consistent estimator is an estimator—a rule for computing estimates of a parameter θ_0—having the property that as the number of data points used increases indefinitely, the resulting sequence of estimates converges in probability to θ_0. This means that the distributions of the estimates become more and more concentrated near the true value of the parameter being estimated, so that the probability of the estimator being arbitrarily close to θ_0 converges to one.

In practice one constructs an estimator as a function of an available sample of size n, and then imagines being able to keep collecting data and expanding the sample *ad infinitum*. In this way one would obtain a sequence of estimates indexed by n, and consistency is a property of what occurs as the sample size "grows to infinity". If the sequence of estimates can be mathematically shown to converge in probability to the true value θ_0, it is called a consistent estimator; otherwise the estimator is said to be inconsistent.

Consistency as defined here is sometimes referred to as *weak consistency*. When we replace convergence in probability with almost sure convergence, then the estimator is said to be *strongly consistent*.

Definition

Loosely speaking, an estimator T_n of parameter θ is said to be consistent, if it converges in probability to the true value of the parameter:

$$\underset{n\to\infty}{\text{plim}}\ T_n = \theta.$$

A more rigorous definition takes into account the fact that θ is actually unknown, and thus the convergence in probability must take place for every possible value of this parameter. Suppose $\{p_\theta: \theta -\Theta\}$ is a family of distributions (the parametric model), and $X^\theta = \{X_1, X_2, \dots : X_i \sim p^\theta\}$ is an infinite sample from the distribution p_θ Let $\{T_n(X^\theta)\}$ be a sequence of estimators for some parameter $g(\theta)$. Usually T_n will be based on the first n observations of a sample. Then this sequence $\{T_n\}$ is said to be (weakly) consistent if

$$\underset{n\to\infty}{\text{plim}}\ T_n(X^\theta) = g(\theta),\ \text{ for all } \theta \in \Theta.$$

This definition uses $g(\theta)$ instead of simply θ, because often one is interested in estimating a certain function or a sub-vector of the underlying parameter. In the next example we estimate the location parameter of the model, but not the scale:

Examples

Sample Mean of a Normal Random Variable : Suppose one has a sequence of observations $\{X_1, X_2, \dots\}$ from a normal $N(\mu, \sigma^2)$ distribution. To estimate μ based on the first n observations, one can use the sample mean: $T_n = (X_1 + \dots + X_n)/n$. This defines a sequence of estimators, indexed by the sample size n.

From the properties of the normal distribution, we know the sampling distribution of this statistic: T_n is itself normally distributed, with mean μ and variance σ^2/n. Equivalently, $(T_n - \mu)/(\sigma/\sqrt{n})$ has a standard normal distribution. Then

$$\Pr[|T_n - \mu| \geq \varepsilon] = \Pr\left[\frac{\sqrt{n}|T_n - \mu|}{\sigma} \geq \sqrt{n}\varepsilon / \sigma\right] = 2\left(1 - \Phi\left(\frac{\sqrt{n}\varepsilon}{\sigma}\right)\right) \to 0$$

as n tends to infinity, for any fixed $\varepsilon > 0$. Therefore, the sequence T_n of sample means is consistent for the population mean μ.

Establishing Consistency

The notion of asymptotic consistency is very close, almost synonymous to the notion of convergence in probability. As such, any theorem, lemma, or property which establishes convergence in probability may be used to prove the consistency. Many such tools exist:

- In order to demonstrate consistency directly from the definition one can use the inequality

$$\Pr[h(T_n - \theta) \geq \varepsilon] \leq \frac{\mathrm{E}[h(T_n - \theta)]}{\varepsilon},$$

  the most common choice for function h being either the absolute value (in which case it is known as Markov inequality), or the quadratic function (respectively Chebychev's inequality).
- Another useful result is the continuous mapping theorem: if T_n is consistent for θ and $g(\cdot)$ is a real-valued function continuous at point θ, then $g(T_n)$ will be consistent for $g(\theta)$:

$$T_n \xrightarrow{p} \theta \quad \Rightarrow \quad g(T_n) \xrightarrow{p} g(\theta)$$

- Slutsky's theorem can be used to combine several different estimators, or an estimator with a non-random covergent sequence. If $T_n \to^p \alpha$, and $S_n \to^p \beta$, then

$$T_n + S_n \xrightarrow{p} \alpha + \beta,$$

$$T_n S_n \xrightarrow{p} \alpha\beta,$$

$$T_n / S_n \xrightarrow{p} \alpha / \beta, \text{ provided that } \beta \neq 0$$

- If estimator T_n is given by an explicit formula, then most likely the formula will employ sums of random variables, and then the law of large numbers can be used: for a sequence $\{X_n\}$ of random variables and under suitable conditions,

$$\frac{1}{n}\sum_{i=1}^{n} g(X_i) \xrightarrow{p} \mathrm{E}[g(X)]$$

- If estimator T_n is defined implicitly, for example as a value that maximizes certain objective function, then a more complicated argument involving stochastic equicontinuity has to be used.

Bias versus Consistency

Unbiased but not Consistent

An estimator can be unbiased but not consistent. For example, for an iid sample {x 1,..., x n} one can use $T(X) = x_1$ as the estimator of the mean E[x]. Note that here the sampling distribution of T is the same as the underlying distribution (for any n, as it ignores all points but the first), so E[$T(X)$] = E[x] and it is unbiased, but it does not converge to any value.

However, if a sequence of estimators is unbiased *and* converges to a value, then it is consistent, as it must converge to the correct value.

Biased but Consistent

Alternatively, an estimator can be biased but consistent. For example if the mean is estimated by $\frac{1}{n}\sum x_i + \frac{1}{n}$ it is biased, but as $n \to \infty$,, it approaches the correct value, and so it is consistent.

Important examples include the sample variance and sample standard deviation. Without Bessel's correction (using the sample size n instead of the degrees of freedom $n - 1$), these are both negatively biased but consistent estimators. With the correction, the unbiased sample variance is unbiased, while the corrected sample standard deviation is still biased, but less so, and both are still consistent: the correction factor converges to 1 as sample size grows.

Gauss–Markov Theorem

In statistics, the Gauss–Markov theorem, named after Carl Friedrich Gauss and Andrey Markov, states that in a linear regression model in which the errors have expectation zero and are uncorrelated and have equal variances, the best linear unbiased estimator (BLUE) of the coefficients is given by the ordinary least squares (OLS) estimator. Here "best" means giving the lowest variance of the estimate, as compared to other unbiased, linear estimates. The errors *need not* be normal, nor independent and identically distributed (only uncorrelated and homoscedastic). The hypothesis that the estimator be unbiased cannot be dropped, since otherwise estimators better than OLS exist.

Proof

Let $\tilde{\beta} = CY$ be another linear estimator of β and let C be given by $(X'X)^{-1}X' + D$, where D is a $k \times n$ nonzero matrix. As we're restricting to *unbiased* estimators, minimum mean squared error implies minimum variance. The goal is therefore to show that such an estimator has a variance no smaller than that of $\hat{\beta}$, the OLS estimator.

The expectation of $\tilde{\beta}$ is:

$$\begin{aligned} E(CY) &= E(((X'X)^{-1}X' + D)(X\beta + \varepsilon)) \\ &= ((X'X)^{-1}X' + D)X\beta + ((X'X)^{-1}X' + D)\underbrace{E(\varepsilon)}_{0} \\ &= (X'X)^{-1}X'X\beta + DX\beta \\ &= (I_k + DX)\beta. \end{aligned}$$

Therefore, $\tilde{\beta}$ is unbiased if and only if $DX = 0$.

The variance of $\tilde{\beta}$ is

$$\begin{aligned} V(\tilde{\beta}) &= V(CY) = CV(Y)C' = \sigma^2 CC' \\ &= \sigma^2((X'X)^{-1}X' + D)(X(X'X)^{-1} + D') \\ &= \sigma^2((X'X)^{-1}X'X(X'X)^{-1} + (X'X)^{-1}X'D' + DX(X'X)^{-1} + DD') \\ &= \sigma^2(X'X)^{-1} + \sigma^2(X'X)^{-1}(\underbrace{DX}_{0})' + \sigma^2 \underbrace{DX}_{0}(X'X)^{-1} + \sigma^2 DD' \\ &= \underbrace{\sigma^2(X'X)^{-1}}_{V(\hat{\beta})} + \sigma^2 DD'. \end{aligned}$$

Since DD' is a positive semidefinite matrix, $V(\tilde{\beta})$ exceeds $V(\hat{\beta})$ by a positive semidefinite matrix.

Generalized Least Squares Estimator

The generalized least squares (GLS) or Aitken estimator extends the Gauss–Markov theorem to the case where the error vector has a non-scalar covariance matrix – the Aitken estimator is also a BLUE.

Bibliography

Acton, F.S.: *Numerical Methods that Usually Work*, Mathematical Association of America, Washington DC, 1990.

Arnold, V.I., Silverman, R.A.: *Ordinary Differential Equations*, MIT Press, Cambridge MA, 1973.

Baez, J. & Dolan, J.: *Higher Category Theory, Contemporary Mathematics*, AMS, Providence, 1998b..

Bell, J. L.: *Toposes and Local Set Theories: An Introduction*, Oxford University Press, Oxford, 1988.

Chance, Beth L.; Rossman, Allan J.: *Investigating Statistical Concepts, Applications, and Methods*, Duxbury Press, 2005.

Dahl, P. F.: *Superconductivity: Its Historical Roots and Development from Mercury to the Ceramic Oxides,* American Institute of Physics, 1992.

David Chandler: *Introduction to Modern Statistical Mechanics,* Oxford University Press, Oxford, 1987.

Eadie, Drijard, James, Roos, and Sadoulet: *Statistical Methods in Experimental Physics*, American Elsevier, 1971.

Eilenberg, S. & Cartan, H.: *Homological Algebra*, Princeton University Press, Princeton, 1956.

Feller, W.: *An Introduction to Probability Theory and its Application,* Wiley, New York, 1968,

Gerber, H.U.: *An Introduction to Mathematical Risk Theory*, Cambridge Univ. Press, Cambridge, 1979.

Greenberg, M.D.: *Advanced Engineering Mathematics,* Prentice Hall, Upper Saddle River, N.J., 1998.

Hatcher, W. S.: *The Logical Foundations of Mathematics*, Pergamon Press, Oxford, 1982.

Hoyland, A., Rausand, M.: *System Reliability Theory: Models and Statistical Methods*, John Wiley, New York, 1994.

Ibach H. and H. Lýÿth: *Solid-State Physics: an Introduction to Principles of Materials Science,* Springer, Berlin; New York, 1996.

Jacobs, B.: *Categorical Logic and Type Theory*, Amsterdam, North Holland, 1999.

Kresin, V. Z. and S. A. Wolf: *Fundamentals of Superconductivity,* Plenum Press, New York, 1990.

Lambek, J. & Scott, P.J.: *Introduction to Higher Order Categorical Logic*, Cambridge University Press, Cambridge, 1986.

Laplace, P. S.: *Introduction to the Analytical Theory of Probability,* Cambridge Univ. Press, Cambridge, 1896.

Majid, S.: *Foundations of Quantum Group Theory*, Cambridge Univ. Press: Cambridge, UK, 1995.

May, J.P.: *A Concise Course in Algebraic Topology*, The University of Chicago Press, Chicago, 1999.

Navidi, William: *Statistics for Engineers and Scientists*, McGraw-Hill, New York, 2006.

Neil Gershenfeld: *The Physics of Information Technology,* Cambridge Univ. Press, Cambridge, 2000.

Olver, F.W.J.: *Asymptotics and Special Functions*, Academic Press, New York, 1974.

Patrick Billingsley: *Probability and Measure*, Toronto, London: John Wiley and Sons, New York, 1979.

Pedicchio, M. C. & Tholen, W.: *Categorical Foundations*, Cambridge University Press, Cambridge, 2004.

Reichenbach, H.: *The Direction of Time*, University of California Press, Berkeley, 1956.

Robert V. Hogg, Allen T. Craig, Joseph W. McKean: *An Introduction to Mathematical Statistics*, Prentice Hall, NY, 2004.

Scott, P. J.: *Some Aspects of Categories in Computer Science*, Amsterdam, North Holland, 2000.

Singh, Simon: *The Code Book : the Science of Secrecy from Ancient Egypt to Quantum Cryptography,* Anchor Books, New York, 1998.

Taylor, P.: *Practical Foundations of Mathematics*, Cambridge University Press, Cambridge, 1999.

Theodore W. Anderson: *An Introduction to Multivariate Statistical Analysis*, Wiley-Interscience, NY, 2003

Unruh, W.G.: *Physics Meets Philosophy at the Planck scale*, Cambridge University Press, Cambridge, 2001.

Vidali, G.: *Superconductivity: The Next Revolution?,* Cambridge University Press, New York, NY, USA, 1993.

Weinberg, S.: *The Quantum Theory of Fields*, Cambridge Univ. Press, Cambridge, 1995.

Index

❑❑❑